The Bonneville Dream

Lorena S. Fisher

Lorena S. Fisher

To Fern

"There is nothing like a dream to create a future."
—Victor Hugo

The Bonneville Dream

Printed in the United States of America

Library of Congress Catalog Card Number: 91-76002

ISBN: 0-8323-0490-5

First Edition 1991

CONTENTS

PROLOGUE

Early in August 1987 the dedication of Bonneville II and the fiftieth anniversary re-dedication of Bonneville I was held. It was a blistering hot Saturday but Bea Abbott and I drove out to attend the festivities because we felt it completed a full circle for us. Bea's husband, Fred, died many years ago and I was very recently widowed, so we went as surrogate old-time engineers. Glenn Meloy of the District Engineers office had arranged a special entrance and parking ticket for us as well as reserved seats. The day before the dedication was "old-timers" day and we heard later that it was a wonderful occasion of reunion, but both of us had other commitments so missed it. When we read the program we were appalled to find *not a single name* of the engineers who built Bonneville I and no mention of the fact that it was built by *The U.S. Army Corps of Engineers*, not the Bonneville Power Administration. Was that intentional or because of ignorance? There was no recognition at all of the Corps! As we mulled over the slight, we became increasingly offended and decided that something should be done about it.

One day I took some family historical items and a box of Bonneville photos and papers to the Oregon

Historical Society. In the course of my conversation with Mr. Tom Vaughn, he mentioned that Al Bauer had also contributed many photos and papers and when I told him I was planning to write a book about Bonneville he offered encouragement. That was all I needed to spur me on.

In June 1988 I invited a group of the old-timers out to my house in Lake Oswego to talk about it. There were eight of us who once lived on the Bonneville Reservation at the same time: Bea Abbott, Ella Torpen, Al Bauer, Dr. Stan Welles, Fern McGee, Roy and Diana Scheufele, and Lorena Fisher. I told the others what I had in mind and asked them to pass on all the stories they could remember. They didn't disappoint me! As time went on I interviewed other old-timers and accumulated quite a wealth of stories, some of them unprintable.

Soon after I started writing I realized that if I expected to get the job done I must have a computer so I invested in one and took a couple of courses. All the other students were young and they probably thought me a nut of some kind to be learning WordPerfect at my age, but I persisted.

So here is the book and I thank all the people who contributed facts and helped in other ways. Pearl Kosta, Reference Librarian of the Lake Oswego Library, has been a constant source of encouragement, and especially helpful in finding information for me.

I dedicate *The Bonneville Dream* to the United States Army Corps of Engineers, who modestly accomplish miracles and make dreams come true.

HOPE

"How would you like to help build the biggest power project in the world?" In 1934 those were electrifying words to a 30-year-old unemployed electrical engineer victimized by the Great Depression. Not that he didn't have plenty of company in that respect. Engineers were a dime a dozen in those days of economic paralysis, especially young ones who hadn't been out of college long enough yet to exactly set the world on fire.

That's the way the letter opened, anyhow, and our friend Wayne Goff, who was already in Bonneville, had enclosed an application blank for Larry to fill in and mail to the U.S. Army Corps of Engineers to let them know he was available. They didn't waste any time in accepting him and he was told to report for work at Bonneville in early January 1935. He was the first electrical engineer on the project but his first job was as an inspector.

Thus began Lawrence T. Fisher's distinguished and exciting 36-year career with the Corps. He retired in 1969 as Chief of the Hydro-Electric Design Branch of the North Pacific Division, Corps of Engineers.

Living quarters were extremely scarce, so our 2-year-old son and I remained in Medford until something, in fact anything, became available. The employees themselves were very comfortable in barracks complete with bathroom and mess hall facilities. The food was even fairly good and the men were congenial.

Housing in the small towns on both sides of the Columbia was completely inadequate for the sudden influx of population. Every shack and hovel even remotely livable was occupied and many of the employees commuted daily from Portland and Hood River. The lucky people were the ones who had arrived soon enough to rent the summer homes of many Portland leading citizens. Of course this was long before the days of mobile homes and other RVs or even TVs, for that matter.

But before I begin my memoirs I think some background data might be desirable.

LEGENDS AND FACTS

Legend has it that Paul Bunyan, the giant of the West, once hitched his blue ox, Babe, to a mammoth plow and set out to do some field work. Maybe Babe got a burr under her tail or a bee in her bonnet or maybe she was just in a rebellious mood that morning, but suddenly she took off and plowed a deep, crooked furrow clear from the mountains to the sea. It's called "The Gorgeous Gorge of the Columbia River."

The facts are far more fascinating than the fiction, for it took the forces of nature (floods, fire, earthquakes, and freezes) millions of years to fashion this fabulous canyon with all its waterfalls, precipices, towering rocks, scenic lookouts, flowers, ferns, forests, submerged forests, and even a natural bridge which eventually fell and formed the Cascade Rapids.

"The Great River of the West" and the mythical Northwest Passage between the Atlantic and the Pacific Oceans had been sought by many nations for three hundred years after Columbus landed on San Salvador in October 1492 on his voyage to find a route to the Indies and the Orient for trading. Several explorers almost found the river, but it was an American, Captain Robert Gray sailing the

Columbia Rediviva who, on May 11, 1792, first sailed across the bar and up the river 11 or 12 miles. (Richard Nokes says 20 or 25 miles.) He named it "Columbia" for his ship, planted the American flag and, by right of discovery, claimed possession of all the territory for the United States. His crew filled the ship's water casks with fresh water from the river and Captain Gray did some fur trading with the Indians for several days. He then returned to the ocean and reported his discovery to the skeptical captain of a passing British ship. Other nations tried many times to discredit the discovery but none ever succeeded. Thus a vast territory was added to the United States.

From *Pacific Graveyard* by James A. Gibbs, Binford & Mort Publishing.

The *Columbia Rediviva* on the Columbia River, May 1792.

Captain Gray, born in Rhode Island in 1755, had gone to sea early in life and was in U.S. naval service during the Revolutionary War. In 1787-1788 he, in command of the *Lady Washington*, and Captain John Kendrick on the *Columbia Rediviva* (both ships Boston-owned) were sailing up and down the Pacific Coast, exploring and trading in furs to take to the Orient where they were highly prized.

Gray and Kendrick didn't get along very well together, being men of entirely different temperaments. Gray was more serious-minded, impatient, obstinate, and irascible, but scrupulous about his duties to his employers and he accused Kendrick of wasting time. Kendrick was inclined to take life easy and he had some grandiose schemes and twisted ideas of honesty in his business affairs. They wintered in Nootka Sound on the west coast of Vancouver Island, Washington. There Kendrick made some repairs on his ship while Gray took short fur-trading trips whenever weather permitted.

The Spanish, who considered the whole west coast theirs, appeared and for several months relations between them and the English explorers were more than a little hectic. When the excitement finally simmered down Kendrick decided to indulge his somewhat puckish sense of humor. On the other hand maybe he was seized by an acute attack of patriotism, being so far away from home and all. Whatever the reason, he staged the first Fourth of July celebration in the Northwest. At dawn he fired a thirteen-gun salute which scared the wits out of everyone — the already-jittery English and Spanish, the Indians, and even the other Americans. The effects were so funny and satisfactory to Kendrick that he repeated the bursts of gunfire off and on all day. Then he topped it all off by inviting the officers of all the ships in the harbor to a feast aboard the *Columbia* to celebrate the thirteenth anniversary of American Independence.

Because the *Columbia Rediviva* was larger they exchanged ships and Gray headed for China with a full cargo of furs. The Chinese especially valued the sable furs. After his cargo was sold, Gray decided to sail on around the world, the first American to do so.

In 1791 he again set out from Boston and arrived in June at Clayoquot Sound in the Northwest. Kendrick was supposed to meet him there, but was delayed by a little trouble with the Spanish over a debt he owed them. Gray proceeded with his fur trading up the coast. When he returned they loaded the *Lady Washington* with furs for the Orient and Kendrick set sail on September 25. He was never seen again.

Gray decided to build a new ship while he wintered in Clayoquot. When Christmas came he arranged a celebration on board his ship, the *Columbia*. It was the first Christmas ever recorded in the Northwest. They decorated the ship and the house which they had built on shore with spruce boughs and invited the Indians to spend the day on board as guests. The chiefs accepted with alacrity and feasted on roast goose and huckleberry pudding. Their squaws would not leave the canoes but they had their fill of food which Gray insisted the chiefs carry down to them.

When spring came the new ship, a forty-five ton sloop christened *Adventure* was launched and sailed north for more furs. Gray took the *Columbia Rediviva* southward with the purpose of really searching for "The Great River of the West." He went as far as Cape Mendocino in California, carefully observing the coastline and trading furs as he went. When the weather became favorable he turned north and sailed to where he had observed a very swift and muddy current on which floated a great deal of land debris. At high tide he sent small boats ahead to tell him depths and find sandbars and reefs. At half sail he cautiously entered what he felt sure was the mouth of the long sought Great River. The three-century-long search had ended with undying fame for Gray.

OVERLAND EXPLORATIONS

Thomas Jefferson had dreamed for twenty years of a nation extending "from sea to shining sea," but his dream was only a dream until he became president of the United States in 1801 and found himself at last in a position to do something about making it a reality. He finally persuaded Congress to appropriate funds for an expedition across the continent, and chose as leader of the long trek Captain Meriweather Lewis, a distinguished military man and famous explorer who at the time was serving as his private secretary. After careful consideration Lewis selected William Clark to be co-commander of the expedition. They studied astronomy, geography and other subjects extensively and made meticulous plans. In 1804 they met and organized their party in Missouri and set out across the continent the next spring. Besides the leaders there were twenty-six soldiers, two interpreters, Sacajawea, the Indian wife of interpreter Charboneau, their baby, and Clark's personal Negro servant, York, who proved to be quite a novelty to the Indians. Jefferson directed them to make painstaking records, which they did. It was a very long journey of incredible hardship and endurance but their detailed records established the

United States' possession of the entire western territory. It also destroyed the myth of a northwest passage to the Orient.

The expedition was supposed to be met at the end of their trek by a ship to take them home, but no ship appeared. A few miles south of the mouth of the Columbia River, at a place later called Fort Clatsop, they built a stockade about fifty feet square and seven log cabins to shelter them through a long, cold winter. With the intermittent help of Indians they managed to find food enough to keep them from starving. In early spring they gave their cabins to the Indians and on March 23 began retracing their steps homeward. The return trip was even more difficult but they finally accomplished it, and found themselves heroes.

Glowing accounts of the expedition triggered the beginning of the "Westward Ho" emigration. First a few hardy pioneers and eventually thousands braved the hardships and sacrifices of the long trip, hoping for better lives in new territory. Miners greedy for gold, devout missionaries dedicated to saving the souls of savages, settlers dreaming of homesteads in a virgin land, and men intent only on adventure and excitement joined the flood of immigrants to the Oregon Country. There were also some anti-social scalawags fleeing from law and justice.

STEAMBOATS ON THE COLUMBIA

Farm folk who settled east of the Cascade Mountains found the soil amazingly fertile and grew crops so bountiful that a problem of marketing them arose. Obviously the Columbia River was the answer, and businessmen who had settled in Portland and formed steamboat companies both increased the number and upgraded the quality of ships to meet the demand. Thus the golden days of steamboating on the Columbia began although there were two places on the river where portaging was required. One of these was about six miles around the Cascade Rapids. The other was about twelve or thirteen miles around Celilo Falls, west of The Dalles. This was accomplished at first by pack mules and Indians, later by mule-drawn wagons on wooden rails covered with strap iron.

Both freight and passenger traffic became so heavy that in 1862 the owners of the Oregon Steam Navigation Company ordered a locomotive for the run around the Cascades. A machinist-engineer by the name of Goffe built it in California and accompanied it when it was sent to Oregon on the steamer *Pacific*. He was hired as engineer on the "pony" which was towed on a barge to Bonneville (at that time only a fish hatchery).

When Goffe decided to give the engine a trial run some of the stockholders insisted on going along for the ride. Goffe told them that they might get dirty but they insisted they didn't care, so the whole crowd, resplendent in top hats and white shirts, climbed aboard and, with a whistle, they started down the track. Everything went fine for a half mile or so until the little locomotive began spitting water and smoke. There was no cover on the cab so they arrived at the end of their trip plastered with dirty water and cinders, coughing and blowing their noses, but jubilant. The Pony was a success! Goffe remarked later, "That was the grimiest crowd of exuberant celebrants I ever saw anywhere."

When he started the return run to Bonneville about three hundred Indians excitedly lined the track. The big chief jumped in front of the engine and yelled, "Hi Ya Skookum Siwash!" He was invited aboard and enjoyed the trip so much that every morning thereafter he was down at the track waiting for his ride.

Two years later the "Oregon Pony" was sent upriver to portage at Celilo Falls. Still later it was sold to a San Franciscan who used it to level sandhills. Then it spent twenty years languishing in a warehouse where it finally burned to a skeleton. In 1905 it was lent for display at the Lewis and Clark Exposition in Portland, was rebuilt, and eventually donated to the State of Oregon. For many years it sat on display in front of the Portland Union Station. Its permanent resting place is in the Cascade Locks Marine Park.

Besides the commercial requirements for transportation there was an ever increasing demand for passenger accommodations which the navigation companies met with upgraded ships. A number of them had Brussels carpets, glittering crystal chandeliers and dining rooms with silver, linen, and crystal. The trip up the Columbia was a popular one for fashionably dressed ladies and gentlemen of Portland. Umatilla House in The Dalles was a symbol of hospitality for steamboat passengers as well as the men who ran the boats.

The Oregon Steam Navigation Company's *Daisy Ainsworth* was perhaps the most luxurious of all. Built by John Holland, master builder of ships, and launched in April 1873, she was considered "queen of the river", but her reign lasted less than four years.

In November 1876, because a herd of 200 cattle was too big for the *Idaho*, she was used as a transport ship and John McNulty, her regular captain, was replaced by her first officer and Pilot Martin Spelling. The *Idaho* made the passenger run. In the *Oregon Historical Quarterly* of March 1933 there is a dramatic account of the catastrophe that ensued. It was a very black and stormy night. Due to mixed-up signal lights on the wharf at the Upper Cascades *Daisy* ran aground and was completely destroyed. Some of the animals swam to land and for many years thereafter they and their progeny were extant in the area. Not very long afterward the pilot died, many thought of a broken heart rather than of tuberculosis, although he was not blamed for the accident.

There were many famous captains in the steamboat era: John C. Ainsworth, Leonard White, Colonel Wright (who was commandant of the military post at The Dalles), R. R. Thompson, L. W. Coe, T. J. Stump, J. W. Troup, E. W. Spencer, Van Pelt, and Sampson — to name just a few. There was even an extremely competent woman captain, Minnie Hill, who worked on the lower Columbia for many years. She was twenty-three when she married Captain Charles Hill in 1883 and in three years she had won an unrestricted pilot's license and the respect of pilot inspectors. Their wedding was "real" but unintended — made legal with the words, "If you're both agreeable to it let it stand". They hauled freight on the 112-foot sternwheeler *Governor Newell* for several years, acquired four more ships, and a son. When they decided he needed more education she moved to a house on land, but whenever one of their boats was short a skipper she "slipped gracefully back into her captain's role ."

The steamboat era flourished until railroads were built and became the favored mode of transportation. There is an interesting account of steamboat days in the March and June 1933, issues of the *Oregon Historical Quarterly*. Nard Jones begins his book, *Swift Flows the River*, with the 1856 massacre at Bonneville. The settlement was left unprotected when most of the soldiers had been sent upriver. The captain of the small steamboat *Mary* made an heroic run to bring them back in time to save some of the settlers.

THE BONNEVILLE DREAM

From the earliest pioneer days people dreamed of possibly capturing at least part of the powerful Columbia River to make life easier. Thus the "Bonneville Dream" was born, when individuals and organizations with imagination first began promoting the idea of harnessing the vast potentialities of the Columbia River. Back in 1874 Congress had authorized the construction of Cascade Locks and Canal to help navigation. It was opened for traffic in 1896 but not finished until 1914. Prior to that, navigation companies had portaged people and cargoes six miles around the Cascade Rapids on their journeys upriver.

Various farmers and orchardists dug deep wells for irrigation while longing to use water from the Columbia.

Other dreamers wished there were some way to make use of all the wasted energy and made tentative plans for power projects but estimated costs were too high for the local financial market. In 1929, when Mr. Franklin Griffith was president of Portland General Electric Company they actually made borings with the hope of building a dam and powerhouse in the Bonneville area but the projected cost was unaffordable and it became just another impossible dream.

At last, in 1925 Congress directed the United States Army Corps of Engineers "to study navigable rivers all over the nation whereon power development appeared feasible and practicable." They were to make general plans for navigation, flood control, water power, and irrigation. A year later the Corps submitted a list on which the Columbia and its main tributaries was stressed as having the most potential. In 1929 the Chief of Engineers authorized more studies, and after consultations with the U.S. Geological Survey, the Bureau of Reclamation, and numerous other specialists the Corps, in its final 1,845 page report, recommended eventual construction of ten dams on the Columbia, with Bonneville the first one downriver and Grand Coulee the farthest upriver.

Meanwhile, the horrendous Great Depression had seized the nation in its clutches. Something dramatic had to be done to get the economy going again and Franklin D. Roosevelt, who was campaigning for his first term as President of the United States, welcomed the opportunity. On one of his trips around the country he visited the proposed site of Bonneville Dam, expressed interest in the power development of the Columbia River, and stated that, if elected, he would authorize the construction of the Bonneville Project as recommended.

After his election in 1932 Oregon's Senator Charles McNary and Representative Charles Martin started skillfully lobbying everyone in sight, and in 1933 Congress voted the sum of twenty million dol-

Ray Atkeson, Photo Art Commercial Studio, *Oregonian*, Northwest section.

Franklin Roosevelt's first campaign visit to Oregon in 1931.

lars to begin construction. This was electrifying news in Oregon! Marshall Dana wrote effusively in the Portland *Oregonian* that it "marked the moment when the U.S. Government caught the Vision of the West and began to make the dreams of its great personalities come true. Began, too, to plant the seeds of those regenerative activities and influences that help to keep government virile and civilizations strong."

Thus, on September 30, 1933, Bonneville Dam became Federal Works Project No. 28 under provisions of the National Industrial Recovery Act. On November 17, when work actually began, plans called for a dam, navigation lock, and a power plant with two generators. Before the two units were completed, however, the Corps was authorized to add eight more, due to estimated power demands. Aristotle wrote many centuries ago, "Hope is a waking dream," and certainly hope had wakened in the Northwest.

Wherever there are dreamers there are doubters who disparage the dreams. So it was with the Bonneville Dream. The National Grange went on record in opposition. *Colliers* magazine published an article, "Dam of Doubt," by one Jim Marshall in which he derided the whole concept. He ended his article by predicting, "There will be some fine concrete monuments scattered up and down the wilderness of the Columbia Gorge, still being paid for by taxpayers." Many prominent Easterners deplored "putting so much money into a project designed to produce so much electricity that might possibly be needed a hundred or more years from now." Some critics called it "Roosevelt's White Elephant" or his "Pipe-Dream." Secretary of the Interior Harold Ickes said it was "a waste of money that could be better used elsewhere." All the pessimistic doubters had to "eat crow" in a very few years.

Now, the dream having become a hope, it was up to the engineers to make the hope a reality. District Engineer Col. C. F. Williams' staff included C. I. Grimm, chief engineer; Ben Torpen, senior construction engineer; and H. G. Gerdes, C. G. Galbraith, R. E. McKenzie, and L. E. Kurtichanof as engineers in charge of dam, powerhouse, lock, and electrical design, respectively. Hired as consultants were prominent engineers in various fields; D. C. Henry and L. C. Hill, who had been consultants on Boulder and Fort Peck dams, were advisors on the main dam and powerhouse. Other nationally known authorities on dam and hydropower were John Hogan, L. F. Harza, F. H. Cothran, J. C. Stevens, and Raymond Davis. Professors Charles Berkey of Columbia University and Edwin Hodge of Oregon State College, distinguished geologists, helped analyze the geological problems and make foundation studies.

Widespread concern about how the salmon would be able to swim upriver to their spawning streams when the project was completed resulted in a fifth objective. Different groups of people proposed various plans for assisting the fish upriver, even including elevators. After careful consideration, a system of fish ladders was built around both ends of the dam and powerhouse. The ladders resemble 40-foot-wide staircases with 16-foot treads and 1-foot risers covered by 6 feet of running water. Submerged openings with controlled jets of water encourage the fish to swim upstream rather than jump. Wickets with white platforms through which the fish must pass were set up, one on each ladder, along with small booths for the people who keep a record via tally boards of the different species as they swim through them. Mary Hill, originally a clerk in the offices, became one of the first "fish counters" and held that job for many years. Among visitors the fish ladders and counting stations were very popular features. People seemed to become so involved that they even cheered the fish on, in their journey upstream.

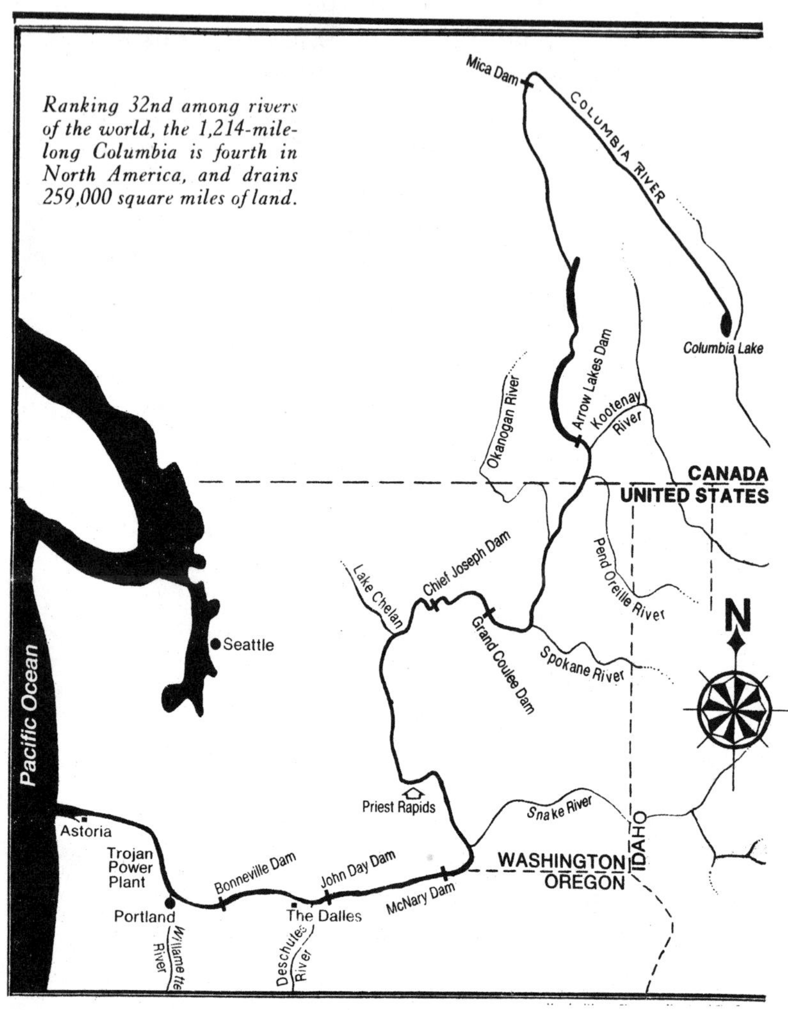

Map by Werner Britner, staff artist of *The Oregonian*

Map of the Columbia and Kootenai Rivers.

Many years latter more sophisticated arrangements made it possible to view them thru glass walls in the Visitors' Center on Bradford Island.

******

It might be historically worth mentioning here that the very first engineering project on the Columbia River dates back to 1871. The Kootenai River rises at the foot of the Canadian Rockies some twelve miles north of where the Columbia appears. It is a much larger and more turbulent stream which flows south, whereas the Columbia flows north along the foot of the Selkirks. They are only about a mile apart. Steamboats plied the Kootenai and a man by the name of W. A. Waillie-Graham, an Englishman and a real dreamer, hatched up the idea of connecting the two rivers by digging a canal between them. He figured that if some of the water of the Kootenai were diverted into the Columbia, boats could go much farther north. Others must have had the same idea before him because the name of the area, Canal Flats, antedated him, but he actually secured permission from the provincial government to dig it. A London syndicate acquired the rights from him and completed the locks in 1888, but according to records no ship ever used them until 1894.

Meantime it overdid the job of draining water from the Kootenai and spring freshets swamped the farms along the Columbia. Every year the canal and locks had to be repaired. Finally in 1893 the farmers on the Columbia, fed up with the floods, went over and built a dam near the Kootenai end. It lasted less than a year. In 1894 Captain F. P. Armstrong, one of the greatest steamboat men of the West managed to get his ship, *North Star*, up the hazardous Kootenai and whistled for entrance to the canal. The people of Canal Flats dashed out, demolished the dam, and operated the locks that lowered the boat nine feet into the Columbia River. The *North Star* navigated on the Columbia for a number of years until it caught fire and burned. Presumably, when the dam was rebuilt this time it remained.

The fact that there are even now genuine ocean salmon land-locked in Kootenai Lake is undoubtedly a result of this engineering project.

TRANSPORTATION

In the 1930s there were three modes of transportation in the Columbia River Gorge. There were steamboats on the river; there were two railways, the Spokane-Portland-Seattle on the Washington side, and the Union Pacific on the Oregon side; and there was a highway on each side of the river.

When the Bonneville Project was being built the only highway on the Oregon side of the river was the famous Columbia Gorge Scenic Highway, another "dream come true."

In 1908 Sam Hill, who at that time was president of Washington State Good Road Association, was commissioned to represent his state at the first International Road Congress in Paris. At his own expense he took with him Samuel Lancaster, a nationally known road construction man, and R. H. Thomson, City Engineer of Seattle. It was Hill's 32nd trip abroad and after the convention they drove by automobile all over Europe, studying roads. Hill spoke German, Italian, and French fluently and knew important people everywhere, including Queen Marie of Rumania and King Alfred of Belgium. The engineers were especially interested in the ancient rock work in the Rhine River country, much of it built in Charlemagne's era. Hill said to Lancaster, "I want you to look closely at those walls, for some day we are going to have similar walls along the bank of the Columbia in Oregon."

Lancaster thought he meant in the far future, but in less than five years after their return home the "Three Sams", Sam Hill, Sam Lancaster, and Sam Jackson, the Portland newspaper tycoon, sat around a table in the popular Chanticleer Inn, east of Portland and about a mile and a half west of Crown Point, and outlined plans for the first great highway in Oregon. They agreed that the highway must enhance the natural beauty and grandeur of the gorge.

In his book, *Lancaster's Road*, Oral Bullard wrote of them, "From these men came the paen of praise, the love song, the poet's dream, that was the Columbia River Scenic Highway."

Chanticleer Inn burned down in 1931 and was not rebuilt.

Oregon did not have its State Highway Commission at that time, but when the plans were made public Simon Benson of Portland was so enthusiastic that he immediately sent $10,000 to Governor Oswald West to get it started. He also bought 300 acres of land including Multnomah and Wahkeena Falls and deeded it to Multnomah County.

Osman Royal donated about two acres at Crown Point and George Shepperd gave eleven acres farther east, Shepperd's Dell. Both Portland newspapers praised the venture and volunteer groups such as the Realty Board and the Portland Ad Club turned out to help get it started. John B. Yeon, "Millionaire Roadmaster" of Portland gave two years of active service without pay, in charge of all road construction except bridges and hard surfacing.

But Lancaster was the inspiration, with the official title of Consultant Engineer. As a young man he had suffered polio and, paralyzed from the neck down, lay in his bed for over a year and a half. Doctors said he would never walk again, but he made up his mind he would some day resume his former active life. By sheer determination and with the encouragement and help of his mother and the girl whom he later married he eventually accomplished his goal.

The scenic highway became his obsession and with an artist's eye he planned all its curves, rock work, bridges, and vistas to enhance the waterfalls and the natural scenery. Stoneworkers were brought over from Italy to build the many miles of dry masonry walls and, quoting Lancaster, "They built their very souls into these walls as they sang their native songs and dreamed of their homeland."

Crown Point had long been considered an unconquerable obstacle to road building, but Lancaster decided to go *over* it and build a viewpoint there. The much photographed Vista House, a small octagonal museum built of stone and glass dedicated to the pioneers, provides a world renowned panorama of the Columbia River Gorge. It was designed by the well-known Portland architect, Edgar M. Lazarus, who echoed Lancaster's sentiment about the site when he said, "The pavilion with its outline against the sky will recall the ancient and mystic Thor's Crown for which the point was originally named."

F. A. Kiser, Portland

Crown Point — Columbia River Highway

The Columbia River Scenic Highway cost 11 million dollars and was paid for mostly by Multnomah, Hood River, and Wasco counties and with some state funds. It was dedicated in 1915 and called by celebrities “a poem in stone” and “one of the great engineering feats of the world.” Tourists from all over the world came to marvel.

I remember the many times my family set out from McMinnville, Oregon, in our Model-T touring car to show the Gorge to visiting relatives from afar. Such trips were high adventure in those days. Most tourists carried picnic lunches and extra tanks of gasoline because gas stations and restaurants were few and far between. On one of our trips a tire “blew out” and we had no spare. My father finally managed to buy one from a not very friendly fellow tourist for four times its value.

At Shepperd’s Dell there was a very tight curve around a monolithic rock which occasionally proved disastrous to big trucks. Once one loaded with asparagus was wrecked and the news went out for people to help themselves. Another time a load of apples was a victim and everyone was invited to help the salvage effort.

One favorite sight to most tourists was Oneonta Canyon. A small stream flows between perpendicular rock cliffs so close together in some places that you can touch both sides with your finger tips. Hundreds of varieties of botanical treasures grow on the rocks: flowers, ferns, lichens, moss, and small shrubs. It is always cool, quiet, and a bit claustrophobic.

Weister Co., Portland

Multnomah Falls

Multnomah Falls has always fascinated tourists. Early in the nineteen century the Oregon Railroad and Navigation Company had a restaurant here so that train travelers could dine while viewing the falls. In 1925 the City of Portland built the Multnomah Falls Lodge as a meal stop for motorists. One of Portland's prestigious architects, A. E. Doyle, designed it to complement the scenery and it is still in operation. Actually it is two falls with cascades between. Sam Lancaster once said to Simon Benson when they were looking at it, "Wouldn't it be nice if there were a bridge across the lower fall with a path up to it?"

Benson wondered how much it would cost, so Lancaster did some figuring on an envelope and gave his friend an estimate. Benson whipped out his checkbook and said, "Then go ahead and build it."

The Bridge of the Gods at Cascade Locks is a beautiful structure that seems almost to float in the air. Geology confirms that the Cascade Rapids were formed when a natural rock bridge across the canyon fell eons ago. The Indians have a romantic myth about which Samuel Lancaster wrote poetically in the *Oregon Journal* of July 20, 1930.

> "Once in the dim and misty past a great stone arch spanned the Columbia like a rainbow between high mountains. Two giant snow-capped male mountains, Mount Adams and Mount Hood, began to fight for the possession of St. Helens, the beautiful maid. This terrible conflict caused all nature to be

> affrighted. Sun, moon, and stars were obscured by dark clouds of ashes. Lightning flashed; both mountains licked out great tongues of fire and began throwing stones at each other. At last Mt. Adams picked up a foothill and threw it at Mt. Hood, but it fell short, landed squarely on top of the stone arch, and broke it down completely. The remains of the fallen bridge formed the Cascade Rapids which is one of the most dangerous stretches of the whole river. Legends tell how when the Indians passed under the "Tomanowos" bridge they lit torches and placed them on the prows of their high-beaked canoes, thus smoking the underside of the arch and proving continuous use 'long time back'."

One of the most spectacular and favorite sights was "The Tunnel of Many Vistas" at Mitchell Point, a few miles west of Hood River. It was a 390-foot-high tunnel with five huge windows cut in the solid rock to provide views of the Washington side of the river. Lancaster's inspiration was the three windowed Axenstrasse in Switzerland. Mitchell Point became a casualty to the freeway built through the Gorge in the late 1950s. Nobody denied the necessity of a freeway, but many people considered the

Gifford, Portland

The Tunnel of Many Vistas, at Mitchell Point, on Columbia River Highway.

tunnel's destruction "wanton travesty and squandered artistic engineering." Certainly the "practical poet," Lancaster, would have found a way to build the new highway without destroying such a treasure.

To provide enough clearance for ocean-going ships to pass under the Bridge of the Gods when the Bonneville Project was finished the Corps of Engineers supervised raising the center section 44 feet. After strengthening and adding to the bridge piers at each end the workers used 500-ton jacks to lift the main section into place, then built new approaches. The Hood River Bridge also had to be raised, but a lift span was installed there.

Over 400,000 cubic yards of rock and other material were removed from the river, but even then the Union Pacific Railroad track had to be raised 35 feet for four miles of extensive tunneling and bridging of Tanner Creek and the Bonneville Fish Hatchery. The Washington railroad also had to be raised to higher ground, as did the highways on both sides of the river.

The Bonneville Fish Hatchery was named for Captain Benjamin L. Eulalie de Bonneville. (He always used the two-syllable French pronunciation.) He was born near Paris, France, in 1793. When he was about nine years old, the family emigrated to the United States and settled in North Carolina. He went to West Point Military Academy, graduated with honors in 1815, and spent fifteen years following with distinction the usual military career, part of it in the Northwest where he became interested in exploring.

He was by nature a bold, daring, and adventurous dreamer, so after his adored wife and little daughter died tragically he took a two-year leave of absence from the army, ostensibly to explore the Northwest for geographical information. Actually he wanted to try his luck as a fur trader and explorer. He kept copious notes of his explorations which he later sold for $1000 to Wahington Irving, the author who wrote the book, *The Adventures of Captain Bonneville.*

He overstayed his leave by almost two years, was believed lost in the wilderness, and was dropped from the army. His dreams of successful fur trading faded when Hudson's Bay Company officials refused to sell or trade assistance of any sort to potential competition.

Eventually he returned to civilization, was reinstated, and returned to active service. Promoted to colonel, he was appointed the first commander of Fort Vancouver when it became a military center for the United States. It was from this assignment that the Bonneville Fish Hatchery received its name and also a mountain peak at the head of Wallowa Lake. Still later he was brevetted a brigadier general and eventually retired in Missouri where he died in 1878.

To make way for the Bonneville Ship Lock the fish hatchery was relocated and modernized. In two or three pools huge sturgeons were kept for visitors to see. Many years later vandals ruthlessly destroyed several of them.

******

Among the famous people who came to see the Columbia River Gorge was Charles A. Lindbergh who did it *his* way, with flair. When he was in Seattle on his personal appearance trip around the country after his famous transatlantic flight in 1927 he decided to detour on his way to Portland and see the Gorge. He cut across to The Dalles and, flying very low and fast, started down the Columbia. The tollgate keeper on the Bridge of the Gods was terrified when he saw a plane heading straight for the three high-tension lines strung across the river not 300 feet from the bridge. Colonel Lindbergh saw the lines just in time to climb over them, swoop down under the bridge, and soar again "as gracefully as a butterfly," in the keeper's own words. Of course the episode was reported and the public loved it.

WEATHER

Weather in the Columbia River Gorge has always been a problem. Winds blow almost constantly. In summer the east wind is hot and dry, in winter icy cold. The west wind is more friendly; cool in summer and warm in winter. Around Crown Point it wasn't unusual in Bonneville days for a car to lose its hood to the strong winds. Winter snowstorms often make the highways impassable, and there is a phenomenon called "tapioca snow" which sometimes pours down canyons in the mountain sides like grain in a chute. In the 1930s it formed drifts across the highway and railroad as deep as fifteen or twenty feet. At that time road equipment was inadequate for clearing it out quickly and prolonged closures weren't unusual. Many of the workers on the Bonneville Project who commuted daily from as far away as Portland and Hood River endured great hardship in winter.

The winter of 1937 was one of the worst in history. The river froze over, and transportation in the Gorge came to a halt. We were living at Warrendale on the bluff across from Beacon Rock. On January 30, although bad weather was threatening and there was already some snow on the ground we dressed up

Fred L. McGee

Old Highway in winter.

very warmly, donned our boots, and drove up to Glendale for dinner and bridge with the Goffs. At about nine o'clock we saw that a blizzard was developing so we headed home.

There were six garages in one building close to the highway above our house for Douglas Lively's tenants. Our's was one of the middle ones. Larry had cleared our driveway before we left, carefully leaving the others open. When we got home we found that our neighbors had cleared their driveways by shoveling snow into our's. Tommy and I left Larry to his shoveling and set out for home, down the steep hill, down six or seven steps, and across the railroad

Fred L. McGee

The big snow of January 31, 1937.

Fred L. McGee

tracks. The ground was frozen so solidly by now that I could barely break the ice with the heels of my boots. Tommy clutched my hand and literally skated on the ice. When we finally reached the steps he said, "Please, Mommy, let's walk over to our road to cross the tracks and then come back home."

That proved to be a good idea, and I thought, "Out of the mouths of babes."

Tommy and I were snowbound for six weeks with snow at least three feet deep. The highway was closed for a week or so and the men had to walk to work via the railroad tracks or the snow-covered highway. There was real concern that the ice would cause serious damage to The Project, and a little cautious use of chemicals was ventured.

At last the warm west wind, called "Chinook," began to blow and within two hours the temperature rose forty degrees. Luckily it was Saturday and Larry was at home. Our front doorsill was slightly below ground level and the snow was melting so fast that we had a stream of water running into the house. Since the ground under the deep snow was still frozen, digging a ditch to divert the water around the corner of the house was almost impossible, but desperation is a great incentive so Larry managed it.

Freezing rain, called "silver thaw," often causes much damage to trees, but it is also one of nature's artists. It transforms the world into a dazzling fairyland. Spray from the waterfalls becomes gossamer veils over trees and shrubs, and great caves of ice form around the falling water. Immense icicles and pillars of ice develop and when the sun shines the whole world scintillates in jewel colors.

When heavy rains fall, to quote one poetic writer, "a hundred unnamed waterfalls spring to life and hang from the cliffs in tiny, silver threads, to vanish when the warm sun dries the uplands."

Average annual precipitation in the Columbia Gorge varies from 42 inches at Portland airport to 100 at Wind River and 150 in the Bull Run area just south of the Gorge. It then drops sharply to 29 inches at Hood River and 14 at The Dalles. Seventy to eighty percent of this occurs in fall and winter; spring and summer are rather dry.

BONNEVILLE — THE REALITY

To get the project started the Portland District of the Corps of Engineers established a five-man Resident Engineer's Office at Bonneville to manage the complex problems of constructing such a gigantic project. The first Resident Engineer was a civilian by the name of George Goodwin who was actually a forestry man. Many people wondered why he was chosen for such an important position in the engineering field. Moreover, he found it difficult to delegate authority and responsibility. The first engineers in the field office were Ben Torpen, Al Bauer, Fred Abbott, E. C. White, and Leo Miller. There was still disagreement in the head office in Portland as to exactly where the dam should be located. A site four miles downriver from Bonneville, Warrendale, was chosen at first but a contract was finally let for the so-called Boat Rock site at the north end of Bradford Island. On November 17, 1933 work was begun on a cofferdam to hold out the water. All was going well until a big flood washed everything out and work stopped on December 25. It was a very discouraging and bleak Christmas Day. Fred Abbott, who argued against this site, couldn't resist saying, "I told you so." One setback does not a failure make, however, so

Start of the first step for a cofferdam at Boat Rock, 1933.

Starting work on the first location at Boat Rock, November 1933. The men in the foreground (left to right): Grimm, Goodwin, Gorlinski, Williams, ___?____.

Site of the present dam, 1933 - 1934.

they picked themselves up and started more tests and studies which produced a new location about 2000 feet downstream, the present site. This change actually resulted in a three-million-dollar savings and a shortening of construction time by one whole working season.

Ben Torpen was experienced in building big dams, having recently returned from Russia where he was an engineer on the great Dnepropetrovsk Dam on the Dnieper River. His experience there, in Peru, and on other projects was invaluable in planning and building the Bonneville Project.

Fred Abbott had been long associated with rivers and harbors projects in the Northwest, so was a "natural" for this one. He was in charge of all inspection, vital to make sure all construction met design requirements.

Al Bauer was widely experienced in gathering data for preparing monthly and final payments for construction contracts so his job was Office Engineer. Mapping the entire area was also one of his responsibilities. This involved determining how much land would have to be acquired for the project itself, for relocating railroads and highways on both sides of the river, and for the huge lake which would form above the dam.

Leo Miller's responsibility was supervising the construction of all the buildings required to house the huge force of workers needed on the project. By the end of January 1934 there were six barracks, each 20 by 40 feet, a kitchen and messhall, and a bathhouse. The number was increased later to seventeen dormitories with adequate baths and mess halls. When employment increased to about 3500 the Corps put up nine ten-man tent houses. They also built a camp for 400 of the contractors' men.

To simplify office work six main construction companies worked under one name, "The Columbia Construction Company." They were (1)Kaiser, (2)Morris-Knutson, (3)Atkinson, (4)J. F. Shea, (5)General, and (6)Utah. Their headquarters were on Bradford Island. Of course there were also smaller companies which were awarded contracts.

******

The Corps of Engineers organized a police force, called the "United States Guards," to protect property, maintain law and order, provide fire protection, direct traffic, and conduct tours for the public. The latter two duties proved the most demanding because over 300,000 visitors came to see the project before December 1935. It took much patience and tact to permit that many people to view it safely without interfering with the work. This was the first federal police force of its kind, ever. A man by the name of Torkelson was the first Captain of the Guards. Harold Leach, who took over later, was rather unpopular because he over-emphasized "spit and polish," discipline, and protocol. "Moose" Clabaugh, a former professional baseball player with the Portland Beavers, followed and was well liked, but "Bonnie" Whitsett was probably the most popular captain of all.

******

Miller's job also included supervising the building of permanent housing for the top personnel. Hollis Johnston, a Portland architect, designed four two-story colonial type frame house plans which were used for the twenty houses. They varied the order in which they built them and reversed the floor plans so they wouldn't look like row houses. They were built on both sides of a circular street near the river around the perimeter of the area chosen for the community. Johnston also designed the permanent administration building and the auditorium used for recreation and meetings. They were of brick, painted white. The basements were poured mostly on ground level and about 700,000 cubic feet of soil was brought in from wherever digging was going on to fill around the walls up to the planned street level. Government workers did this, and private contractors built the houses, which had hardwood floors, electric refrigerators and ranges, and stainless steel sinks in the kitchens, three bedrooms, two bathrooms, dining rooms, living rooms with fireplaces, and central heating. The wood-burning furnaces in the basements were later converted to electricity when the 500 kilowatt "house generator" in the powerhouse was activated. The residences were finished in November of 1934, about six weeks before Larry arrived in Bonneville, and the VIPs joyfully moved in. Some of these were the military officers Gorlinski and Meyers — {young Lt. MacDonald was a bachelor and lived in a barracks building); civilian engineers Goodwin, Torpen, Abbott, Bauer, Miller, White, Laxton, Burke, Rigler, Bray, LePere, and "Ted" Johnson; Dr. Welles; Social Director Orput; Security

Director Torkelson. Originally several of the houses were left unassigned and unoccupied, awaiting the arrival of other engineers as they were needed.

Window shades had not yet been installed and workmen were still painting the houses white and putting finishing touches on them so there were several instances of embarrassing exposure. Tiny, birdlike Mrs. Goodwin, wife of the Resident Engineer, was running around nude one day when she looked up to see a man peering in. Not being very tactful, he waved. Some time later they were introduced and when he said, "I'm glad to meet you," she retorted, "At least this time I have my clothes on." But he had the last word, "I think you were cuter the first time."

The wife of one of the engineers had taken a small room on the first floor of her house as a sewing room. One day she was busy fitting a dress before a full length mirror when she noticed a man staring through the window. He was a gardener who had been weeding a flower bed. She says after that she weeded her own garden beds.

Bad weather delayed the completion of streets, utilities, and landscaping, but in June 1935 everything on "the Reservation" had been completed at a total cost of $402,884.

Landscaping was directed by the general maintenance department with Fred McGee in charge. Joe Elliott, a short, cheerful, and popular Scotsman with a rich Scottish accent which he never lost, was the head gardener for many years. His assistant was a man named Bettendorf and they trained a crew of good gardeners. They planted dozens of quick-growing trees as well as many evergreens on the Reservation, and acres of lawn with automatic sprinklers.

One night a convention of dogs was barking on the big central lawn and Irwin Burke, the cement specialist and director of the concrete laboratory, was trying to sleep. He kept throwing shoes out his window, hoping to make them shut up or go away. Suddenly he remembered that the sprinklers started at midnight, so he rushed out of his house, barefoot and in his pajamas to retrieve his shoes. He didn't escape a soaking, and swore loudly and profanely that someone turned on the water early.

There is another story about him that always intrigued us. Once while his wife was elsewhere a lady visited him and, to make herself comfortable, she removed her corset. (Yes, most ladies did wear them in the 30s.) She had some difficulty getting it back on so to be helpful our hero tried to assist. Eventually he did manage to "stuff her into the thing" and get her presentable.

******

Many rhododendrons and azaleas were planted along the driveway at the entrance, and hundreds of rose bushes farther on. Joe was an expert landscape gardener. There wasn't anything he couldn't tell you about gardens, and many of the residents (including me) learned much from observing him. Each house on the Reservation had a small garden which the tenant was expected to care for. Some were gorgeous and some turned out just so-so. The lawns were always beautiful, however, thanks to Joe and his crews.

Personnel living on the Reservation changed rapidly as the years passed, although some of the first occupants remained until the project was finished.

HEALTH FACILITIES

There were other structures that had to be built. One of the very first, of course, was a first aid station with limited hospital equipment and a doctor's office to replace the original temporary arrangement. The doctor was there primarily for the men on the job, but he also took care of emergencies of the families. A Dr. Kristin was the first doctor at Bonneville, but he stayed only a very short time.

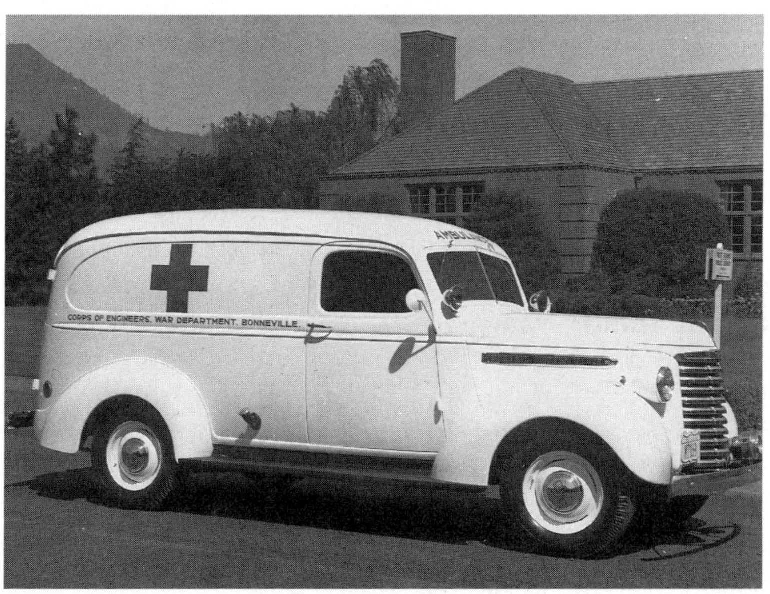

The first doctor in the new quarters was Dr. Stanley Welles, and he tells many stories about extra curricular demands on his time, some of them repeatable. One woman called him to see her little boy, but when he got to her house no child was in sight. She said, "Oh, he's under the bed and I can't get him to come out. Just drag him out by the foot." That being a bit beyond Stan's call of duty, he departed.

Another day he answered the call of a "sick" woman who met him at the door with her robe open and a bottle of liquor in her hand. "Come on in," said she, "I'm just lonely and want someone to talk to." He says he didn't waste any time in making tracks out of there.

He and his wife, Ruth, didn't stay in Bonneville very long. He decided to move to Hood River and establish a private practice there and has lived there ever since. He also had an office in Cascade Locks two or three days a week to accommodate his many friends there and in Bonneville.

When our small son, Tommy, was ready to start kindergarten there was much talk about smallpox vaccination, so we decided we'd better take him to see Stan. That night his temperature started rising and it finally reached 104 the next day. When it climbed another half degree we called Stan and he said, "Bring him up and I'll see what I can do." We were scared silly, but by the time we got to the office the temperature was on the way down.

When I tearfully complained, "He almost died from that vaccination," Stan calmly replied, "Nonsense, but he probably would have if he had contracted smallpox without it." Another year or so later Stan saw him through light cases of chicken pox, whooping cough, and measles, all in a six week period.

Dr. St. Pierre was the next doctor on the project but only for a short time.

Dr. F. B. Freeland succeeded him and to him the Fishers were indebted for saving the life of their dog, an Airedale named Butch, when he was salmon poisoned. Larry had taken me to the hospital in Portland, an emergency trip. When he returned home he found Tommy in tears and Butch half dead with all the symptoms of salmon poisoning. He called "Doc" Freeland and asked if there was any drug that would save him. "Doc" rushed right over, examined him, and said, "He's practically dead now and there's no known cure for salmon poisoning. Let's try a brand new drug, sulfanilamide." The poor dog was so far gone that he didn't even resist when they forced it down his throat. They covered him with a blanket, the doctor went home and my men went to bed hoping but not really believing they would ever again see him alive.

When they opened the kitchen door the next morning he greeted them with tail wagging and a hungry look on his face. I had left a partially cooked pot roast on the stove so Larry seized a knife, cut off a chunk, and offered it to Butch. He gulped it and eagerly waited for more. Obviously the new drug had saved him.

Thereafter we always recommended Dr. Freeland as a top-notch veterinarian. The Freelands owned lovely property on the Deschutes River, called Kahneeta Hot Springs, which they later sold to the Warmsprings Indians who developed it into a very popular resort, complete with a number of tepees for any guest who might prefer them to their regular motel units.

The last man in charge of the first aid station was Vincent Perles, highly esteemed and much loved. He had been a doctor in the Philippines before he emigrated to the United States but settled for the title of First Aid Professional rather than go back to medical school. He remained at Bonneville until the station was closed.

SCHOOL

Another priority building was a grade school. High school pupils went by bus to Stevenson, Washington but there was no place for the younger children. So a very well-equipped grammar school was erected near the highway, and a good faculty was hired. A man by the name of Robertson was the principal. He taught seventh and eighth grades, and "shop" in the basement. Marcia Erickson very skillfully handled the fifth and sixth grades. The only

Reservation, early in construction (lower left), School (lower right).

weak spot on the staff was Mrs. Stevens who had the third and fourth grades. Guyla Galashef, who taught first and second grades was a very talented and personable young woman who ran most of the school programs. She was vivacious and pretty, a real asset in the community. Many of the men, even Captain Des Islets, cast interested glances in her direction but she was careful whom she went out with so the women liked her too. She was the daughter of the Russian man who built the house in Warrendale which the Fishers had occupied for about one and a half years. She also had a brother, Ike, who was one of Joe Elliott's gardeners.

One thing about schools with two grades in a room is worth noting. Often the more intelligent children pick up two grades in one year and a problem arises: should they skip a grade and go on to the next teacher? This happened when our Tommy was in the third grade. He and two other lively little boys, Bobby Des Islets and Gary Blegen, had very high IQ's (Intelligence Quotients were widely used by educators in those days). Bobby was Captain Des Islet's son and Gary was the son of Glen Blegen, one of the best operators in the powerhouse. The Blegens lived in one of Doug Lively's houses at Warrendale, up the hill from us when we lived there. Gary and Tommy had played together since they were three. Tommy was barely old enough for his grade so we didn't want him to skip one although the other parents had already consented.

One day in the spring of 1941 we received a note from the principal asking for an interview. Mr. Robertson said, "I agree with you in principle, but in this case we should make an exception. If we don't, I predict trouble next year. His teacher says he knows the fourth grade subject matter already and will disrupt the class." So he was promoted with his friends.

He says he always found being younger than his classmates in school an embarrassment.

******

Buildings were required by commercial firms which leased them from the government: drug and dry goods stores, cafe, barber shop, beauty shop, movie theater, a temporary reception hall, and even a Safeway store. The historic Bucher dairy just west of Bonneville delivered milk, and various trucks on regular routes from Portland provided us with other dairy products, meat and poultry, fruit, vegetables, and bakery goods.

Jack Gipson was the Bonneville barber and he cut hair for fifty cents. The beauty shop was likewise affordable for many of the women, but most of us opted for "do it yourself" beauty care except for haircuts. I remember sending to Sears Roebuck for home permanents. Very presentable they were, too, but what a hassle!

A man by the name of Rich ran the movie theater.

******

"Bob" White's particular expertise lay in surveying and his was the lengthy job of running hundreds of miles of survey lines to determine how much land would be involved. To start with, the Corps needed to acquire 800 acres below Cascade Locks for the main dam and lock, the temporary and permanent structures, work areas, and railroad and highway relocation. Besides that, some 700 miles of surveying showed that 52,000 more acres would be needed when the big lake filled as far east as The Dalles. All this land had to be surveyed, appraised, and acquired by purchase, grants, and even condemnation. Most of it was simply bought from the owners by the government, but some involved litigation.

The most expensive suit was over an estate owned by Mona Bell Hill. She had been the inamorata of Sam Hill, big, handsome, and one of the most unusual and flamboyant men in Oregon history. They both loved the Columbia Gorge, so Sam bought 32 acres straddling the Columbia River Scenic Highway and in 1928 built a twenty-two room castle on top of a hill for her and their small son. He had it beautifully landscaped to fit into the natural beauty, with many shrubs and trees imported from Japan. The view was magnificent but unfortunately it was situated on the cliff overlooking the ship lock and the relocation of the highway and railroad. By condemnation procedure in February 1934 the Corps of Engineers set a price of $25,000 which she refused. She sued for $110,00 but after much litigation finally settled for $72,500, plus interest.

Mona Bell Hill used the name legally because Sam had arranged a marriage of convenience between her and his cousin Edgar for "considerations." He wanted his son legitimatized as Sam B. Hill, but couldn't do the job himself because of an unbreak-

The Columbian

The 22-room mansion of Mona Bell Hill.

able but long-estranged marriage to a devout Catholic. His wife was the eldest daughter of the railroad tycoon, J. J. Hill, for whom he had worked in Minnesota. The men were not blood relatives. Ben Franklin said, "Where there is marriage without love there will be love without marriage." Sam proved it several times, but he always set up trusts for his "ladies." After severing connection with his father-in-law he had become an entrepreneur deeply interested in good roads and was several times a millionaire on his own. He died in 1931 at St. Vincent Hospital in Portland, Oregon.

Mona Bell and her son lived in the Bonneville mansion for quite a while after the project was

The Columbian

Mona Bell Hill at age 19 in 1909.

begun. I remember little Sam, a handsome but rather lonely child. He was sitting in his mother's big car one day and Tommy and I passed by. He stuck out his hand and said, "Hello, little boy. What's your name?" But Mona Bell just drove away. A Philippine cook, an Indian gardener, and a maid took care of them.

Mona Bell was an interesting and controversial member of the community, a "crack shot" with guns, and a member of the Multnomah County deputy sheriff's reserves. Gregarious by nature, she liked male company but was not very friendly or popular with women. There were even some wayward married men who indiscreetly dared to tryst on the hill. One handsome surveyor visited her often and so did one of the dashing guards. The railroad agent, a woman, was also fond of the same guard. One night at a well-attended dance in the auditorium the two women had a fight over him that almost ended in hair pulling. They made quite a spectacle of themselves.

A real beauty in her youth, Mona left home at eighteen to join William Cody's Wild West Show as a bareback rider. After a couple of years she became bored, started dressing as a man, and for a short time was a bronco rider on the rodeo circuit. After that she took a course in journalism at the University of North Dakota then went to work as a society reporter for a newspaper in Grand Forks, North Dakota. When Sam Hill came to town to lecture about good roads she covered the story and he offered her a ride home.

Sam said later, “She literally swept me off my feet. As big as I am, she picked me up bodily and set me down again across the room.” Through the following years they met occasionally and he always remembered her birthday. In the ’20s she became a crime reporter on a San Francisco newspaper. Eventually they established what might be euphemistically called their “close relationship.”

When swimming the Straits of Juan de Fuca was a fad in Washington State she plastered herself with grease as protection against the cold water, joined the others and almost made it.

When Mona Bell and young Sam left Bonneville she enrolled him in the first of many good private schools in California (eventually he was graduated from Stanford University and became a psychotherapist). He never knew who his father was until he was in his teens.

She went back to Minnesota and somewhere met Osa Johnson of African Safari fame. She was so intrigued by Johnson’s tales of adventure that she set off on a 7,500 mile, six-month safari of her own to Africa without guides or hunters except for natives. In later years she told her son that there was one village in the Congo to which the natives would not take her so she took her cameras and set out alone. When she came near the village she let her hair down and bared her breasts to show she was a woman, sashayed in and began taking pictures. The chief’s wife was very interested in Mona’s “bloomers” and Mona Bell coveted the native’s beaded “G-string” so the two women went into the bushes and traded apparel, to the great amusement of the tribe.

It was illegal to go into the bush without a white hunter so she was arrested, but the authorities let her go because they couldn't decide what to do with her.

After her safari she spent two years traveling around the world and collecting *objets d'art*. She lived the last years of her life in a California nursing home, and died at the age of 91.

******

In the meantime her beloved home in Bonneville had been converted to two apartments for government personnel. Fred and Fern McGee were the first couple to live there, downstairs, then Ollie Moreland and his new wife, Frances Veatch, moved into the upstairs apartment. It was truly a mansion and we women loved to visit the Morelands. Adjacent to the master bedroom there was a huge dressing room walled with slightly concave mirrors that flattered our figures delightfully.

Frances was a lovely, somewhat naive young bride with beautiful hands, anxious to be a good wife. A former music teacher in Hood River, she didn't know much about cooking but wasn't shy about asking for help. Ollie was very fond of my pumpkin pie and angel food cake so he suggested that she get my recipes. She tried the pie first, but it was a failure. She had omitted some items because she didn't have them and used plain milk instead of condensed, so the custard wouldn't thicken. She had added too much water to this first pie crust so it was tough. It was all very disappointing.

After she recovered from that disaster she tried the cake. It didn't rise, ended up about two inches high, and more like a pudding. She brought it over to show me and wailed, "I don't know what I did wrong! I beat it and beat it, but it just shrank and shrank!" When she tried again later she did very well and with practice became a really good cook. In due course Fran and Ollie had two little boys but their marriage ended in tragedy. Fran died of cancer when the younger child was still a baby.

*****

So far as I know, no one ever saw the lower floor apartment. I once heard Fern McGee say, "All the VIPs were dying to see it so I just didn't show it to them."

In October 1934 the Chief Engineer in the Portland office decided to place army officers as administrators of Bonneville. Captain J. Gorlinski became Resident Engineer with Captain Colby Meyers as his Administrative Officer. They and their families moved into houses number 1 and 20 on the Reservation. They brought a new atmosphere to the community, and were very welcome.

Captain Meyers had a rather unpleasant initiation, however. Water for the houses was heated in the furnaces and occasionally very hot water backed up into the plumbing. One unlucky day he flushed the toilet while sitting, and for a few days had to take his meals standing. What a way to start a new job!

Captain Gorlinski had his unfortunate experience at the end of his tour of duty there. He had done such an outstanding job that when the army transferred him to another assignment, the Columbia Construction Company presented him with a fine gun as a token of their esteem. When the Division Office of the Corps in Portland heard about it his gun was confiscated and he was reprimanded for ever accepting it. We ordinary civilians thought his wife should have claimed it as a gift to her and many of us, cynically perhaps, wondered who did get the gun.

Major Theron Weaver and Captain Charles Bonesteele III were the next military officers at Bonneville. Then came Captain R. E. M. Des Islets and Captain "Hab" Elliot. Elliot was there only a year at the most when he overstepped his authority or otherwise displeased the Portland District Office. Colonel Williams unceremoniously yanked him out of Bonneville. Des Islets stayed on until after the war when the Army Administration of Bonneville was transferred back to Portland.

Years later, after very distinguished military careers Meyers, Gorlinski, Weaver, and Bonesteele retired with the rank of general.

******

To handle the ever increasing volume of work, the number of engineers grew steadily. Bill Laxton, Otto Hartman, Howard Rigler, Tom Waring, R. L. Earnheart and many others arrived. One day Torpen, Laxton, and two or three others decided they should go see the source of the Big River, so they drove up to the place high in the Canadian Rocky Mountains where, in the middle of a 25-foot valley of glacial silt they beheld an unimpressive six-foot wide, silently appearing stream. It didn't even have any bubbles. Bill took one look and expressed the feeling of all when he said, disgustedly (to paraphrase his much more vivid word) "Hell, I could *spit* across it without even winding up."

NEW HOME

In March 1935 Larry called me and said, "Come on up. We have rented a house. It's quite primitive but we'll dignify it by calling it our *pièd-à-terre* and keep on looking for something better."

One day on the project he had met a fraternity brother, Al Look, who told him he was moving on to some other job. He was living in "what might be called a house" in Stevenson, Washington. Larry told him to rent it for us, sight unseen. Al told him he doubted we'd like it, but agreed to do so. Al was right; Larry *didn't* like it but we were stuck with it so I packed up what we could bring with us on the train, arranged for the rest of our belongings to be shipped and we headed north. Tommy was ecstatic with his first train ride. We stopped for a brief visit with my father, "Graddy", in McMinnville, then a friend drove us to our new home. The town, Stevenson, got its name from the Stevenson Brothers Logging Company which made its headquarters there. Clark Gable once worked for them as a young man before he started his career as a world famous movie star.

"Primitive" was certainly the right word for our lodgings but it was home, four small square rooms with a toilet partitioned off in one corner, but no hot water or bathing facilities. We did have running cold water in the kitchen sink, with a small, old-fashioned, very steeply slanted drainboard. We just laughed and agreed, "Thank goodness we don't have to carry water from a creek."

In the kitchen there was an old, iron wood-burning stove to cook on and in the front room an equally ancient heater. Nondescript furniture could not really make it a living room but we didn't spend much time there. The "dining room" was furnished with a beat-up table and some chairs. There was a very uncomfortable sagging bed in the other room for us and Tommy slept on a cot. Three or four chests of drawers completed the furniture. Tommy looked around and asked, "Where do we take our baths?"

Since all small boys love the unexpected, I replied, "Wait and see. It's a surprise."

A hole had been bored in the kitchen floor under the sink. The routine was: heat water on the stove, wet him down, soap him, then pour water over him and let it run out on the ground. He thought his baths were fun, but I missed my bubble baths and hot water showers. Larry performed his ablutions at the barracks.

For years I had heard about a "one-armed paper-hanger with the itch," but thought it was just a saying. Our landlord turned out to be just that. Though I'm not sure about the "itch." He chewed tobacco incessantly but when he spit he never missed the stove. He had a buxom, bossy wife who wore the pants in the family. Their name was Ballard. He took a liking to our little boy and came visiting almost every day.

The first morning we were horrified to find Tommy scratching like mad and covered with little red spots that didn't look like measles. "Fleas?" we wondered, but Larry had a hunch. That night after we went to bed he suddenly turned on the light and the terrible truth was out — bedbugs — millions of them. Nothing short of burning the place down could ever get rid of them. We pulled the beds out from the walls, set the legs in pans of water, bought spray, and began a frantic search for new living quarters.

There was an extremely good grocery store not far from our house, owned and operated by a family named Rehal. Young Julius was one of our favorite people and he took a special fancy to our Tommy. He was a brilliant, handsome young man and we were very happy in later years to learn that he had become a doctor, returned to Stevenson and spent his whole career as the most beloved physician in the community. Dr. Julius R. Rehal died on August 23, 1966.

In May Larry finally found an old house in Warrendale, Oregon, across the river from Beacon Rock. He convinced the owner, Douglas Lively, that he should add a bathroom for an additional five dollars a month rent, and we moved in without even one "bedbug hitch hiker" — thanks to a laundry-cleaner. It was a rather "jerrybuilt" two-story house (two small bedrooms over a big living room) but it was *clean*. The kitchen and new bathroom were in a lean-to with a very uneven, you might say "wavy," floor. The cook stove, a tiny old-fashioned one named

"Economy" was just big enough for one pie, and it perched on a little wooden platform about five-inches high and shaped so that the stove was level. There was a big sink with a built-in drainboard, work tables, and enough cupboards to make any woman happy. We didn't own a washing machine but we did have an old-fashioned washboard and wringer so I washed clothes in the bathtub.

There was an enormous space of ground for a garden which we wasted no time planting during the following weeks. As most children do, Tommy found gardening a real adventure and could hardly wait each morning to "run out and see what's growing." When I washed clothes I never knew what treasures I would find in his pockets, "cute little squirmy worms" — his name for them — leaves, pebbles,

Tommy and Tozy.

anything that struck his fancy. We planted all kinds of vegetables in the big garden, and nasturtiums, sweet peas, and hollyhocks around the house.

The first night in our new home we were scared half out of our wits when a Union Pacific train came roaring and whistling by. It took seventeen blasts to get through Bonneville, what with all the roads across the tracks, and the water tower at the station which involved whistles for the station stop, sending the brake man out, calling him back in, and various other signals. The army officers eventually convinced the Union Pacific officials that they could be a little more considerate of sleeping citizens. We knew our house was less than fifty feet from the tracks so it took a little while to get used to the noise.

We finally saved enough money to afford a car, so Larry went to Portland and bought our first automobile, a necessity when you live in the country. We drove our second-hand Chevrolet for several years. We also acquired a dog, a water spaniel which Tommy promptly named "Tozy" after his long time imaginary play mate whom he then promptly forgot.

One day Tozy was barking frantically as he charged the door of an abandoned outhouse. I seized a broom and with Tommy in tow went to investigate. Tozy told us that whatever was bothering him was behind the door so I flailed blindly until we thought it must be subdued. Cautiously we peered in, expecting to see one of the big wharf rats which were a legacy of the old Warrendale cannery that had been located just east of our house in the days of the fishwheels. But it was a tiny cub bear, huddled shivering and trying to protect himself from our

Outhouse where we caught the bear.
Beacon Rock in the background.

ferocious attack. We picked him up, cuddled him, and told him we didn't mean it, hoping his mother wouldn't show up. She didn't, so we had to figure out what to do with him. About that time the neighbor who owned the outhouse joined us and claimed the little bear as his property. Tommy protested the claim "'cause my dog found him," but we bowed out, sure that Mama Bear would come after him during the night. She didn't, but Baby got out of his cage the next morning and headed for the hills with several dogs in hot pursuit. The caretaker of the fox farm just across the highway rescued him and put him in an empty pen. Mama never did come for him so he grew to be a big, beautiful cinnamon bear. To give him exercise Doug Lively, owner of the fox farm, put a collar on him and chained him to a long, overhead wire. Eventually the owner of a Dodson service station bought him but again he escaped. For several years people often encountered him in the woods or raiding "coolers" on back porches for edibles. Most of us didn't own electric refrigerators.

He visited our friends, the Floy Harrells, one night. She taught kindergarten in Bonneville. They lived in a house on Moffett Creek and had been to a dance. They went to bed, but before they got to sleep they heard a commotion on the back porch. He padded into the kitchen, opened the door, and looked out. The bear had unhooked the fastener of the cooler door and removed a sack of apples. When Harrell said, "What the hell do you think you're doing?" he cheerfully answered, "Woof" and kept on munching. Harrell insisted, "Those are *my* apples, so beat it," and moved threateningly. The bear took off, Harrell put the sack of apples back in the cooler and went back to bed.

He was telling his wife about the encounter when she rudely interrupted, "You're drunk again! Sleep it off," but before they got to sleep they heard more noises.

He said, "OK, dear, get up and come with me. See for yourself." This time, when they challenged him, he said, "Woof, Woof" and took the bag of apples with him. He was afraid of no one, seemed to actually enjoy his encounters with people.

Several years later a hunter reported having shot a cinnamon bear wearing a collar. "Sic transit ursa!"

The Warrendale cannery mentioned above was one of many on the Columbia River before fishwheels were outlawed by Oregon in 1926 and by Washington in 1934. The Dodson cannery area, just a mile or two downriver, had grown into a tiny com-

Beacon Rock from Columbia River Highway with the remains of old fishwheels in the foreground.

munity with a store, a post office, and a small restaurant, called Sherman's Inn. During the construction years of Bonneville it was a popular eating place, run by a friendly woman named Mary Sherman. It was still open for business many years later. On a nostalgic trip up the gorge in the early '80s I stopped in one day and found Mrs. Sherman still her amiable self.

Douglas Lively helped with the housing shortage by building about ten small, unpretentious, but comfortable frame houses across the highway from his fox farm and up the hill from our house. All of these houses were still occupied in 1989, so far as I know.

In the spring of 1936 there was an unusual run of smelt in the Columbia and our beach was a good place to "dip" them. Many of our friends came to try the sport. Of course Tommy was an eager helper and of course he fell in. They pulled him out and brought him up to the house bellowing "I'll never go down to that old river again" and he never did, alone.

Our garden produced far beyond expectations, so to take care of the surplus we bought a big pressure cooker and equipment for canning in tin cans. We also canned fruit and, in the fall, elk and venison which Larry hunted in Eastern Oregon. Altogether we had about five-hundred cans of food put away for winter.

Larry was always a very skillful fisherman and often fished for trout in all the lovely streams of the Gorge. I never really enjoyed fishing but always had a license and often met him at the bottom of the falls to claim my limit. I assume it's safe to admit it after all these years. We usually had trout on hand or

A "limit of trout" fromTanner Creek (above the falls).

could easily catch some to provide a treat for friends from Portland who liked to drive out on Sundays to see how the dam was progressing. We had no telephone so in good weather we never knew until they arrived how many we would have with us for dinner, sometimes as many as ten or twelve.

"Gene" (E. C.) Starr came often and we welcomed him. When he was a professor of electrical engineering at Oregon State College back in the 1920s, Larry was a member of what he called his "best class of all," the class of 1929. Most of them became well-known in their profession. Professor Fred McMillan was the head of the department. Gene was a bachelor, only six or eight years older than most of his students, and very egotistical. Putting it in more complimentary words, he was brilliant and he knew it. Whenever the subject of marriage arose he was voluble, listing in detail all the attributes his wife must have. I once heard Mrs. McMillan tell him, "If you ever do find that paragon of perfection she won't give you a second look." (When Bonneville Power Administration was set up he was appointed a consulting engineer and held the job until he died in the late '80s.) We did find him good company, so as the powerhouse took shape and he came out to follow progress he always ate dinner with us. Even when we knew he was coming he was always late, sometimes as much as two hours. I warned him that if he didn't change this habit he would someday arrive and find that we had already eaten. We did just that eventually and put his dinner in the warming oven. He ate it cheerfully but it effected no change. He just explained, "I wouldn't think of coming to your house without freshening up first," even after Larry told him, "We'd rather you come dirty."

Soon after we moved to our "farm" in Warrendale, Nadine Goff invited me to a bridge party to meet some of the engineers' wives. I was only 29 at the time and naturally quite nervous. One of the women was Dolly Miles, a dumpy little lady who had the misfortune of appearing much older than her husband. When we were introduced I bubbled, "Oh, I've heard so much about your son, Jack. It's a pleasure to meet you."

A dead silence fell. Then she replied coldly, "Thank you, but I'm his wife." She never forgave anybody for this *faux pas* and it occurred often.

The Goffs lived in a nice little house Wayne had built on land leased from a Portland businessman, with the provision that the house would revert to the landowner when the Bonneville project was finished. Many others had done likewise. The Linton brothers, Frank and George (safety engineers), built houses next door to the Goffs. When Wayne later decided to join the Coast Guard they sold their house and furniture to us for less than two hundred dollars. (Many years later he retired as a distinguished high echelon officer.) The house was twenty by twenty-four feet in size, but big enough for two people. We converted the small back porch into a room for Tommy. He always liked lots of fresh air so the windows on three sides just suited him. Before we got the screens on them a big cat (not a wild one, thank goodness) jumped in and scared him awake, but he threw it back out the window and Tozy took over.

Tozy tangled with a porcupine soon after we moved there. He came in one day with about eighteen quills in his muzzle. We held him down and pulled them out with pliers. In a few days he showed up with a painful mustache again and we removed them. The third or fourth time it happened there were so many of them that we had to anesthetize him. It took three men more than an hour to get them all out. We found out the hard way that if a dog doesn't learn to avoid porcupines after one or two encounters he never will, so we had to find a new home for him.

We had lived there in Glendale for a year or more, when we were notified that we were eligible for a white house on the Reservation. We sold our little house to Van and Mary Baldwin and moved down with our meager possessions. We drove into Portland to buy new furniture. At the Olds Wortman and King store we found what we wanted on sale. For the dining room we bought a handsome solid black walnut refectory table, six chairs, and buffet which still grace my dining room. The price was $200. We found a sofa and chair, and a few other good pieces. We settled on a 9 x 12 rug for the living room, ridiculously small for the space, but it was a gesture anyway. We rattled around in our big house like walnuts in an apple box, but it was heaven compared to all our domiciles of the previous five years.

We bought an Airedale puppy which grew to be an eighty-four pound "character," everybody's friend. Butch's favorite pastime was fetching rocks for people to throw. With time on their hands, the guards at the entrance to the project found this

Our house on the Reservation.

amusing and kept him busy. His name even appeared in the official log book when a guard with poor aim broke a streetlight and recorded, "Broke street lamp throwing rocks for Butch." More than once they foiled abduction attempts by strangers.

Another favorite dog on the reservation was a wirehair terrier named Pat. He belonged to the Abbotts who had trained him to wipe his feet before entering the house. A feisty little fellow, he challenged any strange pooch that dared to put a paw on what he considered his territory. The Torpens acquired a cocker spaniel named Ebenezer, and when Pat looked up the street and saw him he came prancing stiff-legged to confront him. Eb was hooked up to the clothesline and couldn't get close to Pat. Both dogs were barking frantically and Pat was feeling pretty safe, but he got a little too daring and Eb grabbed his ear. Pat headed for home howling with his torn and bleeding ear.

Bob and Aileen White had two beautiful matched Irish Setters and the Lewises owned the dumbest cocker spaniel you could imagine, but they loved him.

There were also many other animal pets on the reservation. The Laxtons had two Siamese cats that had the run of the house, even going so far as to sleep in the kitchen cupboards. Sometimes you would even find cat hair in your food.

The Karl Druses were gentle people who believed in freedom for their two teenage daughters. Most of us thought one of them went a bit too far when she started cooking her own meals in her room. They were good girls, though, so maybe their parents knew best.

The Holzgangs had a lovely little girl named Betty whom all the little boys hung around. Earl and Verrelle Moody, who lived in the Columbia Construction Company community on Bradford Island, also had a little blond daughter, Sherry, who was a magnet for the youngest generation and Tommy's first love. Jean Lewis, daughter of Frank and Vic, was another popular little one.

Tommy was seven-years-old now and insisted on having a paper route. His first one was up on the highway, but that ended rather abruptly when he was almost run over, bicycle and all, on the high bluff above Tanner Creek. He was then given the route on the Reservation which he and his dog Butch covered faithfully every day. When winter came and the snow was deep Larry couldn't stand seeing him struggling so he went out to help. Tommy smiled his lovely little boy smile and said, "Thanks a lot, Dad, but I would have made it."

All in all, the Bonneville Reservation was a wonderful place to rear children. The guards provided the safety everyone looks for, and the environment was ideal. The teenagers had plenty of talented leadership available for their activities and all the "role models" anyone could wish for.

In later years, after the project was finished and maintenance personnel took over this was still true. For example, Neil Peer's wife, "Boots," was advisor for a group which for several years ran a popcorn and concession booth, located first near the powerhouse and later on Bradford Island when it was developed into a tourist center. Boots told me recently that in the summers at the height of the tourist season they sold hundreds of pounds of popcorn. The young people paid for a bowling alley in the auditorium.

SOCIAL LIFE

Social life on a construction project in the 1930s was a "do it yourself" effort. Every family had a radio, of course, but otherwise people had to arrange their own entertainment. That was before TV had become the principle fascination and movies were about the only ready-made diversion. There was a movie theater in the community for a while and Cascade Locks had one. Fishing in all the beautiful creeks in the area was a favorite pastime of many in the daytime and on weekends, but until the auditorium was built there were very few organized activities.

The athletically-oriented women in the Bonneville area did organize some groups: calisthenics, volleyball, badminton, and other games. Fern McGee, Ella Torpen, Opal Levac, Olive Meyers, and Ruth Carlton were leaders. Others involved were Lynna Lautman, Velda Skinner, Claudine Sporseen, Ruth Hughes, Nadine Goff and many others. They met at first in a rather small old building (formerly a church) with a splintery floor and no heat. This being before the modern colorful athletic suits, the ladies donned their "long johns" to keep from freezing on cold days.

The auditorium and Reservation, upper section.

There was a piano in the building which the "girls" one day decided to move. They huffed and puffed and were pushing it along when one wheel dropped into a hole in the floor and the poor piano fell over. It took three or four men to rescue it. When the auditorium was finished the ladies gratefully moved in.

Don Orput was named Recreation Director and he set about his job with enthusiasm. Everything from athletic events to church services was included on the agenda. Many weekends included a dance on Saturday night and a church service on Sunday morning in the same place. Before the dancers went home they set up the chairs for church.

Programs for special days were arranged, with different groups participating. One of the outstanding ones was a Christmas event one year when the school presented a rather elaborate program. But the finale was a solo by Van Baldwin, one of the employees on the project. He was a member of Chet Duncan's highly talented and popular Portland men's chorus and he chose to sing on this occasion "Oh, Holy Night." It was so beautiful that the audience lapsed into absolute silence and tears glistened in many eyes.

******

"Bridge" was a favorite game at Bonneville, whether "party" or "duplicate." Thursdays were devoted to the latter in the auditorium. Ben Torpen and Jack Miles, at one time ranked as champions in state tournaments, were generally conceded to be top ranking in Bonneville. Sam Gordon, the author of the "Horse Sense System" of bridge, and his son, Don, came out often from Portland to play and sometimes to conduct classes. At first the men and women played together in the auditorium. Larry and I were still living in Warrendale and we had not joined the group because we had no one to stay with our two-year-old son. Ollie Moreland, a young bachelor about our age who often ate dinner with us came up with a solution. He was taking an extension course from University of Washington at the time and said, "Why don't I come to dinner every Thursday evening, stay with Tommy, and do my studying while you two go play bridge?"

We agreed with alacrity and the first time we played was an experience both traumatic and encouraging. Not having played duplicate before, we were so nervous that we could remember for weeks every hand we played. Our last opponents of the evening were Torpen and Miles, the "hot shots." We were quite sure they were saying to themselves, "Ah, a couple of green peas! Here's where we take four 'highs'." They happened to hold very good cards but they overbid on three of the four hands. We were good defensive players and by setting them took three highs and one middle score. After that we had to play North-South with the so called "stronger players", an unexpected boost for our confidence.

Eventually the number of players increased to the point where it became necessary to divide the group. The men continued to meet in the auditorium and Larry's partner was Frank Lewis, the head electrical engineer. The women met in the private homes on the reservation and my partner was "Pud" Cochrane, the wife of the hydraulics laboratory director.

One night Evelyn Laxton was our hostess and one of her duties was to make sure we had "even" tables. She called Clara Rigler and asked, "Aren't you coming to bridge tonight?"

Clara said, "I think I'll stay home tonight unless you really *need* another player."

We *didn't* need her but Evelyn was afraid her feelings would be hurt if she told her to stay home so she said, "Oh, come on over."

When Clara arrived and found she was "extra" she was really angry. "Why did you say you need me when you don't? Edward Everett Horton and I were having a nice visit when you interrupted. It's too late to go back, he's on his way to Portland now, so I'll play bridge and let you sit out."

For the record, Horton was at that time a well-known movie star. Back in the days of the Baker Stock Company in Portland he and Clara had been fellow actors.

There were many private parties of various types and sizes. Captain Meyers and his wife Olive decided to have a treasure hunt Halloween party one year. This is Ella Torpen's account of it. "I told Ben I would rather not go because I had a hunch I'd draw Edgar Kaiser as my partner and he has a reputation as a wild driver. Rumor has it that his wife and another engineer's wife usually sit on the floor of the car when they ride with him because they don't want to see where they are going. Well, we did go and I was right. Edgar was my partner and I never spent such a hair-raising evening in my life, before or after. Clues led everywhere including the railroad station. When I couldn't take any more I suggested that he just ease across the field behind the auditorium and take a shortcut home. I told him nobody would know who made the tracks. That sounded like fun to Edgar so he 'eased across' but drove on over the dam instead of going home. Eventually we did get home and Edgar said, 'Ella, I want you to know that I think you are a damn good sport.' I just went and looked in mirror to see if my hair had turned white. Everyone *did* know who made the tracks because Edgar was the only one crazy enough to do such a thing."

Another year Don Orput organized a community Halloween party in the auditorium with all the usual refreshments and games. The next day someone noticed that a lot of cider and doughnuts were left over and got the bright idea of inviting all the school children for a treat after school on Monday. Bea Abbott says she was there with the other ladies to supervise. After a while the kids were getting very noisy and out of control; yelling and screeching like banshees, falling down and sliding across the floor, fighting, so they sent them home. An adult had a drink of the cider and discovered it had turned "hard." The kids were all higher than kites.

******

One day there was a small private bridge party scheduled and I decided to wear a new dress I had bought in Portland. It caused a minor sensation because it was a maternity dress. We had never intended to raise Tommy as an only child, but the Depression changed a lot of people's plans. He was already seven years old. Our friends rejoiced with us as we looked forward to September 1940. I had plenty of company, for we found that several other couples were in the process of adding members to their families. Among them were the Dements who were anticipating in January.

******

One of the most beautiful women in our community was something of an enigma to everyone. Some of the men thought the women were envious of her. She had been heard to say "I couldn't live without men around" but insisted that their marriage was "entirely platonic." Since she was flamboyant and dramatic in everything she did and said, about the only comment made was "Oh, sure!" One day she announced at another bridge party that she had miscarried one of twins. We were too astonished and polite to say or do anything but gasp or raise eyebrows. After all, she was one of us and good-hearted. One small voice did half whisper, "What next?"

As for me, I must admit that the next time I saw my gynecologist I asked him if such a thing was possible. His answer was, "possible, perhaps, but not very probable." Well, it turned out to be a fact, and in due time their son appeared without any more histrionics. One of the women remarked with a sigh, "No one else I know could have had that happen."

******

On September 20th the Fishers went through the most heartbreaking experience of their lives. Their perfect, beautiful son was born, gasped a few times and died while doctors and nurses worked frantically to save his life. I was moved out of the maternity ward into a general room. There I tried to believe that the unbelievable had actually happened and at last came to a decision. There must be a reason for this tragedy. This child was to have been our last, but there had to be another who was fated to be born. I would not accept defeat, there would be another baby as soon as possible. After a couple of miscarriages, Jack arrived on March 1, 1942 (by

Caesarean Section for the sake of safety) and proved to be all any parents could wish for, beautiful, happy, and good-natured.

The Dements welcomed their daughter, Gail, on January 10, 1940, a tiny baby who developed pneumonia before they could take her home. To save her life, Dr. Bilderback, the famous Portland pediatrician, resorted to a new drug, sulfanilamide, and her parents finally could take her home with them after weeks in the hospital. It soon became apparent that she was deaf as a result either of the high fever she had run or perhaps the drug. She was very tiny, but developed normally in every other way. When she grew old enough for school she was enrolled in a school for the deaf in Salem and received an excellent education. When she completed her education she was employed by Bonneville Power Administration in Portland as a computer data processor, has received several awards for excellence in her work, and has traveled widely by herself.

******

At about this time a small epidemic of adoption developed in Bonneville. Van and Mary Baldwin led off by adopting a beautiful little red-haired baby girl whom they named Nancy. They later added another little redhead, Susie, to their family. Siem and Marge Groth adopted twins, brother and sister. Len and Helen Tucker decided this adoption business was a good idea and made a two-year-old-boy, Johnny, a member of their family.

But the Tuckers ran into trouble when they found out their marriage had never been recorded. This called for a party, of course. The Fishers arranged a big "wedding" for them and invited all the Tuckers' friends. "Bob" Des Islets, the big, good-looking resident army captain, dressed (almost, with the aid of safety pins) in one of his wife's evening gowns and a veil, was the bride; George Hyde, an engineer in the hydraulics lab and the smallest man we could find, was the groom; Bill Laxton in red chiffon and a frizzy wig was the matron of honor; and Larry Fisher in logger's garb and carrying a shotgun was father of the bride. Laxton was late. He took one look at himself in the mirror, got cold feet, and his family had to pour a few drinks into him before he would play his part. It was an impressive, hilarious wedding,

Mock wedding.

punctuated by a small fire in the pocket of Ollie Moreland's new suit where he had stashed his lighted pipe.

******

Occasionally someone had a more formal dinner party and for these occasions a woman by the name of Mrs. Wade was always the indispensable assistant. The Weavers, Colonel Theron and Catherine, and the Abbotts gave beautiful formal parties. Mrs. Wade was considered such a necessary help that once when her husband was due for dismissal from the project Catherine Weaver told her husband he had to find a job for him so they wouldn't move away. He became a member of the guard force and they stayed.

The Torpens once went "all out" and served squab. The next morning when Ella greeted Ermacille Des Islets she was met with cold silence. Later in the day Ella was telling Evelyn Laxton about the snub and asked "What in the world have I done wrong?"

Evelyn replied, "You just committed an unpardonable sin, that's all. You *roasted* the squab. In the South they boil them." The wife of an army officer, Ermacille was also an "army brat," the daughter of an army doctor and reared mostly in the South.

Larry and I were close friends of the Des Islets. They had more or less "absorbed" us as their assistant hosts and I must in all fairness say that we acquired from them a certain "army polish" that proved helpful the rest of our lives. We found some of their habits absurd, for instance pouring their

"Scotch" into "Pinch" bottles and there was one incident that I found quite disgusting. I was in their kitchen helping when a repulsive looking tan bug scurried across the counter in front of me. I said, "What is *that*?"

Ermacille casually answered, "Oh, that's a cockroach. We're having a devilish time getting rid of them. They moved in on us at our previous base." I gulped and tried to swallow my feelings.

******

One day Myrtle Bauer arrived at a private women's party so upset she was practically babbling. "I had a most horrible experience this morning! I was down in the basement doing our laundry. (This was before automatic washers.) For years it has been my custom to add the clothes I am wearing to the last load. I had just tossed them into the washer when there was a knock on the back door, it opened, and a male voice called, 'Just checking the furnaces this morning.' There I stood, naked as a jaybird. I fled into the fruit room and closed the door just as he got to the bottom of the stairs. Panic seized me! What if I couldn't open the door from the inside? Should I try before he left just in case? While I was trying to decide, he started back upstairs and I shot out of my prison just as the back door closed. I tore up to my bedroom and jumped hastily into some clothes. I'm still shaking. One thing is sure, though, I've changed an old habit!"

******

It was generally agreed that the party to end all parties was one the Des Islets hosted. As usual, they invited army people from Vancouver Barracks as well as many Bonneville residents. A hot east wind had been blowing all day and the air was full of smoke from forest fires. Larry and I had taken our house guests, a former employer and his wife, down to Wahkena Falls for a picnic in a cool place during the day but we came home in time for the Des Islets' cocktail party that evening. Bob was one of those sadists who come along when you aren't looking and replenish your drink. Someone said to Ella Torpen, "Come over here and meet my friend, Vivian." Ella did, a bit unsteadily, and said, "I'm very glad to meet you."

Vivian, who was hanging onto the mantel of the fireplace for dear life because she didn't dare turn loose, glared at her and replied, "I don't like you. You speak too much German to suit me and I'm going to have you investigated tomorrow."

"But I don't even *understand* German, much less *speak* it," protested Ella.

"Oh, yes you do," retorted Vivian, "and you're going to be investigated."

I became so warm in the house that I sneaked outside and sat down on the lawn to cool off but some dogs started barking at me and I was accused of barking at the dogs.

Practically every person at that party had imbibed too much before we gave up and went home to bed. As Al Bauer says, "There never was another Des Islets."

******

John Des Islets, Bob's very suave older brother, brought his wife and four children to visit Bonneville one summer. John was graduated from West Point in the top of his class before Bob, but left the army to assume command of a military academy in the east. (When the United States entered World War II, John immediately returned to active duty and was a general by the end of the war.) Larry went out and caught enough trout to feed everyone and we invited all the Des Islets over for a very informal dinner. Every time I got up from the table to go to the kitchen for something John would get up too. Finally I said, "I don't care what army etiquette says. When I get up will *you* please stay put? You embarrass me."

The construction companies had an extensive and rather close social structure of their own in addition to the community one. Most of their engineers lived in houses on Bradford Island. Earl and Verrelle Moody were close friends of ours with whom we kept in touch for many years after we left Bonneville.

The only really lavish wedding to occur at Bonneville was that of Harriet Burke and young Lieutenant Bob MacDonald soon after the white houses were finished. This was Bob's first assignment as an army engineer and he was finishing his training under Captain Gorlinski. Since he was unmarried he lived in the bachelor accommodations. Harriet was very attractive and one of the few eligible young women on the reservation, and as if that weren't enough, her mother encouraged their courtship. Women are accused of being gossips but some of the engineers said he practically lived at the Burke house and "never had a chance." The nuptials were solemnized in a Catholic Church in Portland with only the family and very close friends in attendance, but the reception was an all out extravaganza in Bonneville. The marriage proved to be a happy one.

******

On October 30, 1938 Orson Welles touched even *our* lives. His realistic radio account of the invasion by Martians caused a flurry of frightened people to load up their cars and seek advise from their employers as to where to go and what to do. Captain Des Islets dismissed them thus, "Go on back home and go to bed. It was just a radio show."

******

In 1939 Frank Bradley, chief of the finance section, decided there should be a softball team in Bonneville, so we all became softball enthusiasts. He talked it up among the young men and organized them in to a champion group. He was their coach. Larry Fleskes, shortstop, recently gave me a list of the players. The pitcher was Biff Jorgenson; catcher, Roland Miller; first base, Bud Matlock (who later married Bradley's beautiful daughter, Gloria); second base, Morrie Stremich; third base, Jim Reimer. Other players were Neil Peer, Walter Schlafe, Bob Willis, and Stan Danielson. This team played Portland teams of that era and won most of their games.

Larry Fleskes was our son Tommy's hero. I was not very well that August so in lieu of a big birthday party for Tom we told him he could invite one boy to have dinner with us, expecting it would be Bobby Des Islets with whom he played most often. He said he wanted Larry Fleskes, so I told him he could also invite one boy his age. He picked Bobby Elliot, the son of the other military man on the reservation. We decided we simply could not omit Bobby Des Islets so we had dinner for six. I always had a feeling that my seven-year-old son had outwitted me!

When World War II was declared most of the softball team joined the armed forces. Fleskes served with distinction on the aircraft carrier, *U.S.S. Bismarck Sea*. Many years later, in 1958, he was elected to the Portland Softball Hall of Fame. Thirty years after that, in 1988, he was inducted into the Northwest Hall of Fame.

******

There were many off-beat hobbies among the employees on the Bonneville Project. Dick and Peggy Earnheart never seriously tried to play bridge. Peggy suffered the effects of "undulant fever" and was never very well. They had no children. Dick has always suffered from recurring attacks of curiosity which usually resulted in interesting activities. During his career he has invented several useful items. While he lived at Bonneville he had a very powerful microscope which was intriguing to many of the engineers as well as others. Dick recently told me the following story which he swears is true.

One couple, close friends of theirs who also had no offspring, decided that they should conduct some experiments to find out why. So after dinner one evening the guests retired to a bedroom to produce the necessary substance for the experiment. The results looked "encouraging" and the hopeful couple did later have their family.

CASCADE LOCKS

Cascade Locks is an old, historic settlement. It was named for the ship locks and canal, opened to traffic in 1896 and completed in 1914, which lifted ships over the unnavigable rapids. In 1905, the year of the Lewis and Clark Exposition in Portland, 1,417 ships used it, carrying 133,070 passengers, mostly sightseers. It is the only incorporated town between Troutdale to the west and Hood River to the east, although there are many small settlements with many longtime residents. The Bonneville Project brought a welcome influx of construction people and a boom to business. All available living quarters were very soon occupied, with long waiting lists of would-be tenants. The main street's name was Wa-Na-Pa until the highway was built.

The town was quite politically oriented. When the dam was started the mayor was Charlie Nelson, an opinionated, irascible individual. The huge increase in population rendered the water system totally inadequate so the city council finally decided to build a new one. After a big squabble and much argument in the council, bonds were sold and work started. Contractor Henry Kuckenberg, who was building it, was progressing nicely until the

Cat Creek, the original water supply for Bonneville.

time for payment came. The mayor refused flatly to sign the necessary warrants because he considered it too expensive and elaborate for the town. He knew that legally he didn't have a leg to stand on, but he was obdurate. Finally some of the leading citizens started a recall campaign against him and two councilmen, Max Millsap and Carl Epping. The editor of the local newspaper was a good friend of Nelson but he wrote in favor of the recall because the water system was needed, the bonds had been issued and the work begun, and it was only logical to take whatever steps were required to finish it. Nelson was recalled by a big majority in the election but the two councilmen remained in office.

The editor was Hugh A. Scott, who came from Portland in 1936 to help Jack Travis of Hood River establish a newspaper *The Bonneville Dam Chronicle*. Its slogan was "the best newspaper in the world by a damsite." As far as we could tell, Scott was as near a one-man newspaper as could be found. He was everything from janitor to editor. Travis showed up once in a while, but he was also trying to start a newspaper in Hood River and left the *Dam Chronicle* to Scott. According to Scott's own account, he was working at the Portland *Oregonian* when he heard about the Bonneville job and wasn't keen on the idea of moving from the city to the "sticks" until he found out that E. P. Hoyt, managing editor of the *Oregonian*, thought young Hugh might get some good, practical experience. Scott's father, Quincy Scott, was the *Oregonian*'s cartoonist at the time and Hoyt did not favor nepotism. Young Scott moved out to Cascade Locks and roomed with the operators of the local Standard service station. Their names were Lew and Berta Morgan. He shared a bathroom and breakfast with the family. Lew was a tough veteran of World War I, and he told Hugh about being in a Siberian hospital with a gangrenous foot which the doctors told him had to be amputated. He had his gun under his pillow and when they came to take him to surgery he pulled it out and told them he'd blow their heads off if they tried to operate. They changed their minds, the foot healed, and Lew was still walking eighteen years later.

In an article he wrote for the *Oregon Historical Quarterly*, Hugh tells an interesting story about his news-gathering adventures. He was over on Bradford Island one day talking to the men in the office of the Columbia Construction Company. He showed interest in their slide rules and wondered "if they could figure out how many cubic feet are in a cubic light year." They said it was "a cinch" and he left. A week later he was making his usual news-gathering rounds when the secretary told him Edgar Kaiser wanted to see him. With some trepidation he walked into his office and, without even a greeting, Kaiser barked at him, "Are you the person (he used an unquotable phrase) who asked my engineers about cubic feet and light years?"

Scott gulped and replied, "Yes sir, I'm afraid I am."

"Then get the hell out of here and never come back. My engineers have done nothing all week but mess with that stupid problem, and they keep coming up with different answers."

Even though Hugh was "scared" he thought it was all very funny and cautiously covered his same "beat" the next week on schedule. Nothing happened but Scott admitted he did rather avoid another encounter with Kaiser.

J. B. Laber, a Portland businessman, owned a lot of property in Cascade Locks and was in favor of anything that would increase property values if it didn't cost him anything. Every month Laber came out to collect rents and gather information about town events. He wasn't very popular but he was a frequent visitor and wielded a certain amount of influence. After he died his daughter Maureen, a very well known educator in Portland, took his place.

A man by the name of Cooper had a stable of horses which he rented to people who liked to ride. Margaret Bourke White, a photographer for *Life* magazine once came to Bonneville to take some pictures for an article. Ella Torpen and Al Bonesteele were driving her around to see the Gorge when they passed Cooper's sign and White said she enjoyed riding horses. So they went in, chose likely looking mounts and started out. Some time later when they thought everything was going well Margaret's horse suddenly dumped her flat on her back. The other women were terrified, thinking, "What have we done? Have we killed her?" But White picked herself up, climbed back on her horse, and assured them she was OK.

At one time there had been a sawmill at the east end of town, but all that was left of it was a huge pile of sawdust for which there was no known use back in the 1930s. It caught fire and smoldered intermittently for several years, never dangerously and usually unnoticeably. Nowadays, when we pay high prices for sawdust for our yardwork, it's painful to contemplate the huge waste.

Not far from it was the Methodist Church and well-founded rumor had it that almost next door was a popular "house of ill repute," but *maybe* it was just rumor.

Penn's Tavern was a motel and restaurant of good reputation. Our friends, Clark and Isabel Bogart, lived there for a while until they found a house on Moffett Creek. The tavern was not elaborate, but it was clean, comfortable, and served good food.

Merrill's Tavern was another popular and respectable restaurant which served good food. They had a dance floor, and "canned" music, sometimes even a live band. The "big snow" of 1936 caused the roof to collapse and it was not rebuilt.

On a side street Ernie Manchester had a prosperous truckline which hauled goods between Portland and points as far east as The Dalles. He also had one of the most gorgeous daughters around.

The tallest building in town was the three-story Lakeside Hotel, notorious for its paper-thin walls. On the two upper floors there were about two dozen barely furnished rooms. Never having visited the place, I leave to the reader's imagination what sanitary facilities were provided. Many risque stories were told about it.

The post office and several small businesses were housed in a two-story building, the second floor of which was used by a lodge or some such organization. They had old-fashioned dances there and the music included western, folk tunes, polkas, and even square dancing. A Doctor Johnson had his office in the building and he was often wakened to sew up somebody's head after a free-for-all at Erickson's Tavern or one of the other hangouts of the rougher construction men on Saturday nights when brawling was a favorite way to let off steam. There is a story that Erickson's was the "stomping ground" of big, burly "Tiny" Mitchell, whose idea of fun was "cold-decking" as many men as possible, whether the violence was deserved or not. Eventually the men got tired of his playfulness, ganged up on him, and beat him to a pulp. He was an amiable citizen after that and fitted right in with the crowd.

Noble Hyde ran a men's store, and was a peaceable person, but his son Bobby was something else. He fancied himself a real man about town; a cornetist, a bandleader, a dancer, a comedian, and a general hell-raiser. He even grew marijuana on the roof of his father's garage and not only admitted it, but bragged about it. Back then it wasn't illegal, of course, but he would have fit right into the 1980s drug society.

There was a well-patronized movie theater in town and, down by the Locks, a beautiful and well-tended little park which invited much picnicking.

Between Cascade Locks and Bonneville, on Eagle Creek, was a ranger station. Al Weisendanger, the "Keep Oregon Green" man, was resident forester and he was a very popular man. On the crest above the creek was a settlement called Eagle Creek Heights. It was a community of comfortable small houses occupied by various office personnel and other workers. Among them were Jim and Ann Bell, and Charlie and Charlotte Johnson, long-time friends of ours.

Farther west, across the highway from the entrance to the dam project, Cook's Addition was home for many more employees. Morrie and Kareen Vennewitz and a number of other people had built houses on leased land in back of Cook's.

"Johnny" Walker had a small dance band which played for some of the local dances. Driving out from Portland, he once lost the hood of his automobile to the raging winds at Crown Point.

Hal Babbitt was a personable young man who once owned a small restaurant in the area. He was an artist who made and gave the Abbotts a lovely picture of their little dog, Pat. Bea still has it hanging in an honored place in her house in Beaverton. In later years Edgar Kaiser commissioned him to help decorate his estate in the San Juan Islands in Washington state.

Going on west from Bonneville there were communities called Shady Nook, Glendale, Wee Cove, Bonnie Vue, and Bonnie Villa, all gobbled up when the superhighway was built in later years. Bonnie Villa was a very pleasant tavern and sandwich shop run by Rolf Enquist which provided good but not exactly gourmet food.

At Latourell Falls there was a two-story house whose owners took pride in arranging luncheon or dinner parties given by Bonneville people who lived in houses too small for large groups. In the 1850s there was a town on the river called Latourell where sternwheelers stopped with passengers who danced outdoors to the music of the Latourell Band.

Multnomah Falls has always had a fine restaurant.

CONCRETE LABORATORY

Because of the complexity of the project a special concrete laboratory was established. Irwin Burke, who had the reputation of being possibly the best in the field, was named the director and he had a crew of experts working with him. They made few mistakes.

The amount of concrete required to build the Bonneville Project is almost beyond imagination. If it were loaded onto freight cars the train would extend over 200 miles, or from Portland to Seattle. Over one million cubic yards were used, but the problem was made more difficult by the many different structural requirements. The concrete laboratory developed seven different mixes for the various needs of the spillway dam. In addition to these, the powerhouse, navigation lock, and other structures required fourteen other blends. Corps inspectors kept constant, careful watch over every stage of the process, from start to finish. Work on the spillway was begun by the Columbia Construction Company in June 1934.

But before any concrete could be poured, engineers had another problem to solve — how to divert the current of the river. Because of summer

Moving a big dipper dredge into position to dredge for crib footing at south end of main dam.

Pouring concrete — Bottom–dump bucket in upper right.

floods the work season was only about eight months long, from approximately August to March. After many hydraulic studies which included weather and time they decided to build huge timber cofferdams "tailored" to fit the riverbed and anchored to bed rock.

They were designed by George Gerdes, chief engineer for the main dam, and required eight million board feet of timber (much of it 12 x 12's) at a cost of $2.5 million. At that time it was the largest cofferdam job ever attempted and attracted wide attention.

The first cofferdam enclosed the south half of the spillway site and partial construction of the dam was accomplished in the low water period of 1935-36. The cofferdam was then removed to permit the river to flow between and over the top of the piers while another cofferdam was placed in the north half of the river. After the completion of the whole north half of the spillway during the 1936-37 season, the builders placed a prefabricated structural steel cofferdam over the top portions of the unfinished south half and brought them to total height.

Once the cofferdam problem was solved, it was time to pour concrete. The crews, 10 to 20 men each, placed the concrete by bottom-dump buckets of 8 cubic yards capacity hauled to the proper places by two cableways of 25-ton rating. Then they used vibrators to puddle and compact the concrete.

In September 1935 the Corps set up a special structural steel inspection unit to ensure the highest standards of work. To quote one of the engineers, "Bonneville Dam is no ordinary rivet-tapping job: tolerances are limited in certain instances to 0.01 inches in 5.0 feet... Then, in order to expedite opera-

Early volunteer "inspector".

tions, it has been necessary to carry on construction simultaneously with design and detail." Because all important steel connections were welded, all welders were tested to standardize the quality of work. Standard tools varied in precision so special devices were designed to co-ordinate all measuring tools. A 15-man inspection team made up of engineers and technicians expert in welding and measuring covered the dam site around the clock and the powerhouse 16 hours a day.

Early volunteer “inspector”.

Concrete pouring was an exciting procedure to watch and it never stopped, day or night. The gigantic equipment, the bright lights, the orderly activity, and even the incessant noise never ceased to astound observers. Lois Myers, a reporter on *The Portland News-Telegram* wrote about it in November 1935 after seeing it at night.

— Her account —

> "Lights everywhere, great floods of light that make an oasis of brilliance in The darkness of the night. In the background, dimly visible, black masses of the steadfast mountains, undisturbed by this noisy confusion made by puny men, moving like busy ants about the depths of the river's forsaken channel. Nearly 100 feet below sea level they pour tonight — monster buckets of concrete dangling from twin high-lines, swiftly carried, carefully lowered to find the precise spot in the excavation for which they are intended. A bucket emptied of its 16-ton burden swings over to the north bank for a reload. As it descends, a toy-like train runs out on its track to meet it. Swiftly a chute leaps out from the car, a cataract of concrete flows into the yawning bucket and... another contribution is on its way to the building of Bonneville Dam."

Even in daylight visitors were fascinated by the magnitude of the equipment and the drama of creating such an incredibly colossal structure. We who lived there never ceased to be impressed.

On September 1934 the first concrete for the foundation of the powerhouse was poured and by June 1935 about ninety percent of the substructure had been finished. In October 1935 the Corps let contracts for the superstructure and for the design and manufacture of turbines, generators, and other electrical components. Now Frank Lewis, Larry Fisher, and Richard Earnheart, the three electrical engineers, took over and did what they were hired for... produce power.

Starting to concrete-in pier eight of powerhouse

Steel workers placing reinforcing steel on downstream side of intake piers, powerhouse and sub–structure. Note: No hard hats.

"Hard hats" were unheard of in those days, but accidents were few. The men all wore whatever head gear seemed comfortable to them or worked bareheaded. For example, in order to be easily found on the project Larry Fisher always wore a French beret, a navy blue Pendleton style wool shirt, a bright yellow necktie, and khaki pants. Many other engineers also wore distinctive apparel on the construction sites.

Speaking of accidents, I vividly remember one bad one that occurred in the powerhouse. Three men were working at the bottom of one of the turbine holes, 90 feet down, when someone on top accidentally tripped a switch that released carbon dioxide used for fighting fire. The sirens and alarms were screaming and all the workers ran for gas masks, but the three men at the bottom were trapped. One big, burly young man, Freddie Schneider by name, managed to get to the top on his own. With no mask, Larry Fisher raced to the bottom and helped another of the men halfway up and masked men took them both to safety. The third man died. Larry couldn't speak above a whisper for several weeks but recovered except for intensification of his emphysema.

******

A few more statistics about the project might be interesting and enlightening at this point.

Thirty million board feet of timber were used in building the dam, for cofferdams to shut out the river while excavating for bedrock, for foundations, forms in which to pour concrete, and many other uses.

To increase the capacity of the spillway dam to pass large streamflows during flood months, the Corps widened the channel for three miles upstream from 800 to 1200 feet by blasting and removing 954,293 cubic yards of material to other places. The spillway can pass a flood 27 percent greater than the maximum level recorded in 1894, the one that inundated Portland.

The length of the powerhouse is 1,027 feet, width and height 190 feet each. It houses 10 generators weighing 900 tons each with a total generating capacity of 518,000 kilowatts.

Superstructure of powerhouse.

All ten generators in service.

Looking east in lock chamber.

Looking west in lock chamber.

The navigation lock chamber is 500 feet long and 76 feet wide with a maximum lift of 70 feet. The electrically-powered silicon steel miter gates at the upper end are 45 feet high and those at the lower end 102 feet (as tall as a ten-story building). The downstream gates weigh 525 tons each. To enable small river craft to avoid dangerous moorings at low stages of the river there are six floating mooring bits in the lock walls, designed by Brigadier General John Kingman.

The number of employees on the government payroll at a give time reflected the progress of the construction. When the project was fully underway the work force averaged about 3000, with skilled workers earning a minimum hourly wage of $1.20 and unskilled $.50. (Remember the Depression?)

Trimming the rock walls in the locks.

The *Wheeler* was the first ship through the locks in August 1937 (before the project dedication).

HYDRAULICS LABORATORY

Soon after the project was begun a small hydraulics laboratory was established on the bank of the river, not far from the concrete laboratory. Paul Heslop was the director and J. C. Stevens was the consultant. It was the job of these people to build scale models to ascertain the feasibility of plans. The first ones were a 1 to 36 scale model of the three spillway gates and a 1 to 100 scale model of the river from the dam to the head of the Cascade Rapids. It was much less expensive to build models than to build structures and wait to see what would happen.

In January 1939 the Corps expanded it to about fourteen men with R. B. "Bax" Cochrane as director and George E. Hyde his assistant. Both of these engineers were transferred from a Mississippi laboratory where the same type of work was being done. The personnel included Bill Dement, Carl Anderson, Alton Alspaugh, "Bud" Matlock, "Red" Holiday, Larry Metcalf, Orville Johnson, William Gustavson, Harry Theus, and Al Chanda. Bob Grimm was official photographer, and Paul Shephard the clerk. Somewhat later Wayne Buchanan replaced Grimm.

The plan was for the lab to operate for a few more years at Bonneville then be relocated to Mississippi but it remained for a much longer time, even after the construction men left and the maintenance staff took over. When "Bax" Cochcrane was transferred to Portland, Marvin Webster, newly returned from a Harza Engineering Company assignment in Pakistan, became director. Many of the engineers in the Portland District office thought that George Hyde should have received the appointment, believing him to be the man for the position.

Hyde resigned from the Corps of Engineers, moved to New York to work for a private company for a few years, but returned to the Corps when he was offered a position in Dallas, Texas, and later in the Portland District office.

As time went by the engineers and workers finished their jobs and were transferred to Portland, Walla Walla, and other places where the Corps of Engineers was constructing other projects.

The exciting days of Bonneville construction were over.

WAR

The first Sunday in December 1941 was one that probably no one who was living at the time can't recall exactly where he was and what he was doing. We had as houseguests Mary and Cormack Boucher of Seattle, Washington. Larry and Cormack, both electrical engineers, were graduated from Oregon State College in the same class. That Sunday morning soon after breakfast they went over to the powerhouse to see how things were going. Siem Groth was the operator on duty, and they found him almost incoherent. He had the radio on and told them to listen. The broadcaster was saying, "I repeat, early this morning the Japanese bombed Pearl Harbor, Hawaii. We do not yet have a complete report on exactly what happened, but please stand by and we will tell you as soon as possible. We do know that the results are catastrophic. This undoubtedly means *war*."

Larry and Mac came home, told us the news, and turned our radio on. As the day passed the news became worse. The United States was in deep and totally unexpected trouble. For months the Japanese had been given free access to the whole Bonneville Project, and they were suspiciously eager to photograph everything, even including all the blueprints. The engineers on the job had protested, but were overruled and told they were much too suspicious. Literally hundreds of Japanese, many with two or three cameras each, had been going over things with a disquieting zeal, but someone in the Portland office was heard to remark flippantly, "Maybe the little bastards will get some education." The engineers at Bonneville were afraid they had gotten too much.

That very day about two hundred soldiers were sent out from Fort Vancouver to patrol and protect all the railroad tracks, high ways, and bridges in the area. The U.S. Coast Guard patrolled the river. A .50 caliber machine gun was mounted inside the power house pointed at the main door. The Corps of Engineers built concrete "pill boxes" at the entrance to the project and on the Washington side of the spillway. They were manned around the clock by armed guards of the augmented Federal Bonneville police force.

In spite of Siem Groth's protests Federal authorities immediately moved him and his family to Portland "for their own protection." He had emigrated from Germany very soon after World War I and was a naturalized U.S. citizen of many years standing, but his mother and twin brother still lived in Germany.

That Sunday night was a long one of sleepless anxiety. We could hear airplanes flying very high. No one seemed to know whose planes they were and we never did find out. The dogs barked incessantly.

More than just a few fearful families fled to Portland.

Early Monday morning all workmen available were set to work camouflaging everything. The dam, powerhouse, and lock were spraypainted blue gray to make them harder to see from the air. All the houses and other buildings were painted green, as were all the streets. In less than two weeks Bonneville had been completely disguised and was actually a military establishment. Many of the younger men went to Portland and volunteered for military service. Ollie Moreland qualified for lieutenant in the Navy, but George Hyde enlisted as a private in the Army when he failed the eye test. He immediately started the procedure to become an officer, and eventually retired from the Army Reserves as a lieutenant colonel.

Larry Fisher had been in ROTC in college and decided to enlist in the Navy. He was accepted as a lieutenant commander on one condition — that the Army Corps of Engineers release him from his commitment to them. This they refused to do because they considered him indispensable to their effort to put a maximum amount of power on the lines as quickly as possible. This they did so effectively that in a very short time there was sufficient electricity for three aluminum plants in the area to produce enough aluminum for 50,000 airplanes and to power the shipyards on Swan Island in Portland and in Vancouver, Washington. Henry Kaiser's shipyard launched a Liberty Ship per day for a prolonged period of time, no doubt partly because of the skill of about a thousand ship carpenters who had been trained at Bonneville in building the hull-shaped draft tubes in the generators. Now they built ships.

In place of the military service denied him, Larry joined the Coast Guard Reserve for duty in the Portland area and until the war ended he spent all his weekends and holidays with them on patrol duty.

An *Oregonian* headline of June 22, 1942 reads,"JAPS LOB SHELLS ON VANCOUVER ISLAND," and on June 23, "FORT STEVENS TARGET OF JAP SUBMARINE," but that was all the public ever heard about it. Balloon bombs made their appearance, launched haphazardly from Japanese submarines with the hope of setting forest fires to divert attention and efforts of the military. There was speculation as to why there were so many small fires in Oregon forests, but even though one bomb exploded and killed five or six people they were never publicized. Military public relations preferred not to broadcast news of that sort, with the purpose of keeping the Japanese government uninformed as to the effects of any attacks they might make on the mainland.

Gradually, as it became evident that Bonneville was not in imminent danger of being bombed, everyone settled down to a new, more serious way of living. Everyone played less and worked more. Although I was having some health problems, I signed up with the Portland Red Cross knitting group and organized a large circle of knitters in Bonneville. We were proud of the fact that when we took them a box of items no one ever questioned whether they were acceptable because they always were. I personally taught at least twenty women to knit.

Many women volunteered assistance in other ways. The Red Cross Bloodmobile came out frequently for blood donations and received help with the bookwork and publicity. They usually set their operation up in the auditorium. One day when a large group of workmen came in to donate blood, one big, burly fellow fainted dead away. That started an epidemic of "passing out" until the nurses sent them all away and told them to come back later.

Someone decided that we should grow "Victory Gardens" and a committee was set up to arrange for them. Just west of the white houses there was a large unused area of land which seemed a good location and plots were laid out for all who wanted them. Many of us planted vegetables there, more or less successfully. Rationing of various foods and especially gasoline became part of the American way of living.

One day in the early spring of 1943 I received an anonymous letter berating me for "neglecting your poor sick father who is now in the McMinnville hospital." Our baby, Jack, was only a year old and we had wondered why "Graddy" wouldn't come to see us. He seldom wrote letters, and he always said that he had to earn a living and he didn't have enough gasoline to drive that far.

"Pud" Cochrane always loved Jack, so she offered to "baby-sit" while I drove down to see what the situation was. I had barely enough gasoline to make the trip and planned to bring Dad home with me, but the hospital flatly refused to dismiss him. They were treating him for "fibrillation of the heart." His clothes were hanging in the closet and I knew he would go home the first chance he got, so they promised to hide his clothes and keep him there. I had just arrived home when I got an angry call from the hospital saying he had left and they wanted nothing more to do with him. Somehow we bought enough gasoline for another trip and I went down again early the next morning. I asked him why he hadn't told us he was ill and he said, "Because I knew you'd take me home with you. I love all of you, but I want to stay here."

This time I had to take care of closing his house and other business. I stayed overnight and found that someone had stolen his large chest of expensive carpenter's tools and other possessions. He tried to put too big a chunk of wood in his little stove and almost set fire to the house. I decided to stop in Portland on our way home for him to see a heart specialist. Dr. Harold Gillis prescribed digitalis, complete rest, and an enriched diet to regain lost weight. We continued our trip to Bonneville. Very confused, unhappy, and homesick, he refused to take his pills and eat properly. He was with us for about six or seven weeks before he died. It was a heart-rending period and it took a long time for me to get over it. Of course I had feelings of guilt for not knowing sooner about his illness.

FINIS

By 1943 the ranks of the engineers who built Bonneville were thinning out rapidly by transfer to other jobs or by resignation. The operational and maintenance personnel were taking over, and electrical engineers were the last to leave. Dick Earnheart was transferred to Umatilla and built a house in Hermiston where he has lived ever since. "Doc" Lewis and Larry Fisher were transferred to the Portland office.

In the summer of that year Larry went to Portland to find a place for us to live. Gasoline was still in short supply, so he took a room in a hotel downtown and began contacting real estate people. We were sure we didn't want to live on the east side of the city on account of the weather. After three or four days of searching he put a down payment on a house in Eastmoreland. It was eighteen years old and somewhat run down, but he couldn't find anything for rent any better and he liked the area.

When he walked into the powerhouse the next morning he met a new operator on the job, Andy Junor, who said, "I heard last week you are looking for a house in Portland. Would you like to rent mine?" He named a quite reasonable rent and told him its location, which was only a few blocks from the one we had the down payment on. Larry was never a violent man, but he said he never felt more like making mayhem. He just asked, "Why didn't you tell me last week?" Of course we could have kissed our $200 goodbye, but that was a lot of money in those days and we decided to stick with our deal, even though Andy's house was much better. We hated to leave Bonneville and felt that perhaps we would never again live in so beautiful a place but in late summer we made the move in time for Tommy to begin the school year at Duniway.

Realizing that the city was no place to take Butch, our beloved dog, we had made arrangements for the army to take him into the "canine corps," but just before we moved they announced a delay. "Could we please keep him for two or three months?" So he moved with us and we soon found out there were dozens of dogs in the area and not even any leash laws. Butch went to school with Tom and made friends with all the children. But tragedy struck. Some dog-hater gave him strychnine and he died only a few hours later. Matched white collies, and several other dogs in the neighborhood met the same fate within two weeks at the hands of an unknown person.

Frank and Vic Lewis also bought a house in Eastmoreland. Vic still lives there, thanks to one of her granddaughters who, with her husband, occupies the downstairs apartment. Vic's daughter, Jean, and two other granddaughters still live in London, England.

The Cochranes, too, bought a house a few blocks away from us.

L. E. Kurtichanof and his wife, Nancy, lived for many years in a house behind us. He was the Corps' first consulting engineer on the Bonneville Powerhouse.

Roy and Diana Scheufele bought a place in southwest Portland where they still live and Al Bauer is also a resident of that area.

Ella Torpen lives in Hillsboro, Oregon; Fern McGee Fryer in Woodburn.

The Dements, who had been living in one of the apartments converted from the former barracks of construction days, bought a house in east Portland and Bill commuted to Bonneville. In the late 1950s Bill Dement was still working at Bonneville when the Corps of Engineers decided to investigate how the cement under water was lasting. The photographers on the site were not capable of taking the pictures needed.

Bill loaded himself with cameras and other equipment and was lowered into the depths of the powerhouse to photograph the draft tubes and turbine blades, the navigation locks, the spillways, and other ordinarily underwater areas for pictures of possible cavitation. The assignment was both dangerous and interesting.

About that time the freeway through the Gorge was begun. No one could deny that it was essential to Oregon's progress, but we who spent our Bonneville years traveling the Scenic Highway still sometimes wistfully revisit it. What is left of it is well worth a trip. There is a group of Oregon people who work to preserve the remnants of one of the most beautiful and famous highways ever built in the United States.

The Bonneville we knew is no longer there. The Reservation is gone, the white houses all sold and moved elsewhere to make room for the new and larger ship lock. Recent headlines in newspapers say in their exaggerated way "the dam is wearing out," but it will always remain "The Bonneville Dream" in the hearts of us who made the dream a reality.

Bonneville — the dam, the powerhouse, and the reservation. The concrete lab and the hydraulics lab are in the center. (May 1932)

EPILOGUE

More than fifty-five years have passed since the original Bonneville Project was dedicated, followed by all the other dams and powerhouses on the Columbia River. Lawrence T. Fisher spent his entire career in Oregon with the U.S. Army Corps of Engineers. He retired as Chief of the Hydroelectric Design Branch of the North Pacific Division in 1969. Our two sons were military fighter pilots; Tom, a Marine who flew the A-4 Skyhawk for thirteen months in Korea and another thirteen months in Vietnam (205 missions); Jack, an Air Force pilot, flew the famed "Phantom" F-4E for over twenty-two years. He was stationed in Khorat, Thailand the same year Tom was in Vietnam and flew more that 200 missions. They spent Christmas together at Jack's base. Jack went back later and flew another 100 missions. Both of them were decorated many times, including the Distinguished Flying Cross, but were fortunate enough to miss the Purple Heart.

In 1956 we were lucky enough to find a lot on Lake Oswego just as beautiful as Bonneville and built a modest-sized house which we have occupied ever since. My hobby is growing roses, about 225 bushes. I've won several trophies for my garden.

We "old-timers" who are still living find nostalgic pleasure in returning to Bonneville, although only memories remain of the exciting days of construction.

BIBLIOGRAPHY

Allen, John Elliot. *The Magnificent Gateway*. Forest Grove, OR: Timber Press, 1984.

Brogan, Phil F. *East of the Cascades*. Portland, OR: Binford & Mort, 1965.

Davenport, Marge. *Fabulous Folks of the Old Northwest*. Tigard, OR: Paddlewheel Press, 1986.

Donaldson, I. and Cramer, F. *Fishwheels of the Columbia*. Portland, OR: Binford & Mort, 1971.

Friedman, Ralph. *A Touch of Oregon*. Portland,OR: Pars Publishing Company, 1970.

Haig-Brown, Roderick L. *Return to the River*. New York, NY: Morrow, 1941.

Hanley and Lucia. *Owyhee Trails*. Caldwell, ID: Caxton, 1973.

Holbrook, Stewart H. *The Columbia*. NewYork, NY: Rinehart, 1956.

Holm, Don. "Where Rolls The Columbia." *The Oregonian Northwest Magazine*, 20 July 1975.

Hosmer, Paul. *Now We're Loggin'*. Portland, OR: Binford & Mort, 1930.

Irivng, Washinton. *The Adventures of Captain Bonneville*. Klickitat edition. Portland, OR: Binford & Mort, 1954.

Jackson, W. H. with Ethel Dassow. *Handloggers*. Anchorage, AK: Alaska Northwest Company, 1974.

Jones, Nard. *Swift Flows the River*. New York, NY: Dodd, Mead Co., 1940; Reprint: Portland, OR: Binford & Mort, 1964.

Lancaster, Samuel C. *The Columbia, America's Great Highway Through the Cascade Mountains to the Sea*. Portland, OR: J. K. Gill Co., 1926.

Lewis, Meriwether and Clark, William. *The Journals of the Expedition Under the Command of Captains Lewis and Clark*. New York, NY: Heritage Press, 1962.

Lavender, David. *The Great West*. New York, NY: American Heritage, 1965.

McDonald, Lucile. *Search for the Northwest Passage*. Portland, OR: Binford & Mort, 1958.

Meacham, Walter. *Bonneville the Bold*. Portland, OR: Metropolitan Press, 1934.

Portland General Electric Company. *History of Portland General Electric Company from 1889 to 1981*. Portland, OR: Portland General Electric Company, 1982.

Ross, Alexander. *Adventures of the First Settlers on the Oregon*. London, 1849; Reprint. New York, NY: Citadel Press, 1969.

Scott, Hugh A. "Reminiscence of Bonneville." *Oregon Historical Quarterly* (Fall 1980).

Smith, Courtland L. *Salmon Fishers of the Columbia*. Portland, OR: Portland State University Press, 1979.

Steber, Rick. *Where Rolls the Oregon*. Union, OR: The Bear Wallow, 1985.

Sunset Books. *Rivers of the West*. Menlo Park, CA: Lane Magazine & Book Company, 1974.

Tuhy, John E. *Sam Hill — The Prince of the Castle Nowhere*. Portland, OR: Timber Press, 1983.

Ward, Harriet. *Gold Saga of the Umpqua*. Portland, OR: Metropolitan Press, 1966.

Willingham, William F. *Army Engineers and the Development of Oregon*. Washington, D.C.: Government Printing Office, 1983.

Willingham, William F. *Water Power in the Wilderness*. Portland, OR: U.S. Army Corps of Engineers, Portland District, 1987.

Wilson, Fred W. "Lure of the River." *Oregon Historical Quarterly* (March, June 1933).

INDEX OF NAMES